Rameshwar Adhikari · Santosh Thapa
Editors

# Infectious Diseases and Nanomedicine II

First International Conference (ICIDN-2012),
Dec. 15–18, 2012, Kathmandu, Nepal

Springer

*Editors*
Rameshwar Adhikari
Central Department of Chemistry
Tribhuvan University
Kathmandu
Nepal

Santosh Thapa
Department of Microbiology
Tribhuvan University
Kathmandu
Nepal

ISSN 0065-2598 ISSN 2214-8019 (electronic)
ISBN 978-81-322-1773-2 ISBN 978-81-322-1774-9 (eBook)
DOI 10.1007/978-81-322-1774-9
Springer New Delhi Heidelberg New York Dordrecht London

Library of Congress Control Number: 2014930226



Printed on acid-free paper

Springer is part of Springer Science+Business Media (www.springer.com)

# Preface

This Special Issue of *Advances in Experimental Medicine and Biology* is devoted to the First International Conference on Infectious Diseases and Nanomedicine (ICIDN) which was innovatively conceptualized by the Nepalese Association of Medical Microbiology (NAMM) and the Nepal Polymer Institute (NPI) in association with Kathmandu University, Dhulikhel, Nepal to promote interdisciplinary scientific study for systemic understanding of connections between major human diseases and their treatment regime by applying the tools and techniques of nanotechnology. The conference organized in Park Village Hotel and Resort, Budhanilkantha, Kathmandu from December 15 to 18, 2012, was accompanied by a one-day Short Course on *Recent Trends in Nanomedicine and Infectious Diseases* and had the motto: *Interdisciplinary Collaborative Education for Research & Innovation in Biomedical Sciences*. The ICIDN-2012 was attended by approximately 180 scientists from 21 countries including dozens of eminent scientists from throughout the globe.

One of the aims of this congress was to get researchers from the fields of nanomedicine (including physics, chemistry, and engineering) and various fields of infectious diseases (including pathology, microbiology, and clinical sciences) together at the same congress to explore common ground. The conference focused on themes on nanomedicine with special attention to anticancer therapy, antibiotic resistance, and virology.

The Inauguration Ceremony was attended by Rt. Hon'ble President of Nepal, Dr. Ram Baran Yadav as Chief Guest. In his inaugural address, President Yadav stressed on the role of young scientists and students in developing the nation through science, technology, and innovations. He praised the efforts of the ICIDN organizers to arrange such a global scientific meeting of enormous scientific and practical significance in Nepal. The inauguration ceremony was also addressed by Prof. Geoffrey L. Smith (President of the IUMS; Head of the Department of Pathology, University of Cambridge, England), Prof. Jamboor K. Vishwanatha (Dean, Graduate School of Biomedical Sciences, University of North Texas Health Science Center, USA), Prof. Surendra Raj Kafle (Vice Chancellor of Nepal Academy of Science and Technology (NAST)), Prof. Ishwar Chandra Dutta (Chairman of Tribhuvan University Service Commission), and Dr. Rajendra Koju

(CEO, School of Medical Sciences, Kathmandu University, the representative of Vice Chancellor of the University). The congress was opened formally through the video lecture on *Helicobacter pylori* by Nobel Laureate Prof. Barry J. Marshall from The University of Western Australia.

The conference served as a meeting point for young researchers and students with their peers and well-known scientists for making scientific contacts and exchanging their experiences. The enthusiastic participation of numerous young scientists and students from all over the world led to vivid scientific discussions during the breaks, poster sessions, as well as after the sessions. Ms. Pramila Adhikari, Ms. Manisha Pradhan, and Mr. Eak Dev Khanal were bestowed best Poster Presentation Awards (selected from total 44 presentations) during the closing ceremony. Prof. Geoffrey L. Smith (President of the IUMS) and Prof. Chirika Shova Tamrakar (Dean, Institute of Science and Technology, Tribhuvan University, Kathmandu) expressed their positive impression about the content, quality, and performance of the conference in their closing remarks.

We are thankful to the IUMS delegates including President Prof. Geoffrey L. Smith (Department of Pathology, University of Cambridge, England); President Elect Prof. Yuan Kun Lee (Yong Loo Lin School of Medicine, National University of Singapore, Singapore); Vice Presidents Prof. Stephen A. Lerner (Wayne State University School of Medicine, Michigan, USA) and Prof. Pierre J. Talbot (University of Quebec, Canada and organizing President of IUMS 2014 congress), and IUMS Ambassador to Asia Prof. Fusao Tomita (Hokkaido University, Japan) for their cordial support.

We would like to extend our special thanks to Prof. Suresh Raj Sharma (then Vice Chancellor of Kathmandu University) for the kind patronage of the conference. Particular thanks go to Mr. Dhurba Acharya, Mr. Radium Adhikari, Mr. Santosh Adhikari, Mr. Netra Lal Bhandari (Conference Secretary), Ms. Sulakshana Bista, Mr. Kedar Nath Dhakal, Dr. Surendra Kumar Gautam, Ms. Jyoti Giri, Mr. Khagendra Prakash K. C., Mr. Eak Raj Kadariya, Mr. Suresh Raj Kandel, Mr. Santosh Khanal, Mr. Bhuvan Khatri, Mr. Prayag Raj Kuinkel, Mr. Surendra Kumar Madhup, Mr. Rajesh Pandit (Treasurer), Mr. Tarini Prasad Paneru, Mr. Chakravarty Paudel, Ms. Sharmila Pradhan, Mr. Krishna Kumar Raut, Mr. Din Dayal Regmi, and many other students and researchers affiliated to Kathmandu University, NPI as well as NAMM, and Nanochemistry and Polymer Research Group at Tribhuvan University for their untiring efforts to make ICIDN-2012 a grand success. Thanks are due to the members of National and International Advisory Boards for their valuable feedback and support.

The financial support to the congress was generously granted by the Nepal Polymer Institute (NPI), Kathmandu University, University Grants Commission (UGC), National School of Sciences (NSS), Lainchour; Kantipur College of Medical Sciences (KCMS), Sitapaila, and National Health Care Ltd., Kathmandu.

This conference has definitely sensitized the young scientists and students from Nepal toward strengthening Research and Education in various avenues of nanomedicine, infectious diseases, virology, and antibiotic resistance. The congress is hoped to be a milestone in promoting education and researches in the field of microbiology, drug delivery, and biomedical sciences in particular fostering our endeavors in sustainable applied Nanoscience and Nanotechnology in Nepal.

We take this opportunity to thank Mr. Aninda Bose and Dr. Tobias Wasserman from Springer publication for their interest and cordial support in bringing out this special issue.

Rameshwar Adhikari
Santosh Thapa

# Contents

# Contributors

**Rameshwar Adhikari** Central Department of Chemistry, Tribhuvan University, Kirtipur, Kathmandu, Nepal; Nepal Polymer Institute, P.O. Box 24411, Kathmandu, Nepal

**Ram Prasad Aganja** Central Department of Biotechnology, Tribhuvan University, Kirtipur, Kathmandu, Nepal

**Lucia Baldi** Ecole Polytechnique Federale de Lausanne, CH-1015 Lausanne, Switzerland

**Anastasia Baranova** State Research Center for Applied Microbiology and Biotechnology, Obolensk, Moscow Region, Russia, 142279

**Evgenia Baranova** State Research Center for Applied Microbiology and Biotechnology, Obolensk, Moscow Region, Russia, 142279

**Om Basukala** Central Department of Biotechnology, Tribhuvan University, Kirtipur, Kathmandu, Nepal; Nepal Academy of Science and Technology, Khumaltar, Lalitpur, Nepal

**Sergey F. Biketov** State Research Center for Applied Microbiology and Biotechnology, Obolensk, Moscow Region, Russia, 142279

**Charlotte Boucher** Department of Microbial, Biochemical and Food Biotechnology, University of the Free State, P.O. Box 399, Bloemfontein 9300, South Africa

**Robert Bragg** Department of Microbial, Biochemical and Food Biotechnology, University of the Free State, P.O. Box 399, Bloemfontein 9300, South Africa

**Élodie Brison** Laboratory of Neuroimmunovirology, INRS-Institut Armand-Frappier, Institut national de la recherche scientifique, Université du Québec, 531, Boulevard des Prairies, Laval, QC, Canada

**Ananda M. Chakrabarty** Department of Microbiology and Immunology, University of Illinois College of Medicine, 835, South Wolcott Avenue, Chicago, IL 60612, USA

**Elke Coetsee** Department of Microbial, Biochemical and Food Biotechnology, University of the Free State, P.O. Box 399, Bloemfontein 9300, South Africa

**Marisa Coetzee** Department of Microbial, Biochemical and Food Biotechnology, University of the Free State, P.O. Box 399, Bloemfontein 9300, South Africa

**Marc Desforges** Laboratory of Neuroimmunovirology, INRS-Institut Armand-Frappier, Institut national de la recherche scientifique, Université du Québec, 531, Boulevard des Prairies, Laval, QC, Canada

**Anuradha Dude** Central Drug Research Institute, Lucknow, India

**Ivan A. Dyatlov** State Research Center for Applied Microbiology and Biotechnology, Obolensk, Moscow Region, Russia, 142279

**Galina Feduykina** State Research Center for Applied Microbiology and Biotechnology, Obolensk, Moscow Region, Russia, 142279

**Victoria V. Firstova** State Research Center for Applied Microbiology and Biotechnology, Obolensk, Moscow Region, Russia, 142279

**Andrew Gdowski** Department of Molecular Biology and Immunology, Graduate School of Biomedical Sciences, University of North Texas Health Science Center, Fort Worth, TX 76107, USA; Texas College of Osteopathic Medicine, University of North Texas Health Science Center, Fort Worth, TX 76107, USA

**Aleksey A. Gorbatov** State Research Center for Applied Microbiology and Biotechnology, Obolensk, Moscow Region, Russia, 142279

**Arina Jansen** Department of Microbial, Biochemical and Food Biotechnology, University of the Free State, P.O. Box 399, Bloemfontein 9300, South Africa

**Ramaz Katsarava** Institute of Chemistry and Molecular Engineering, Agricultural University of Georgia, 13 km. David Agmashenebeli Alley, 0131 Tbilisi, Georgia

**Johan L. F. Kock** Department of Microbial, Biochemical and Food Biotechnology, University of the Free State, P.O. Box 399, Bloemfontein 9300, South Africa

**Liubov Kolombet** State Research Center for Applied Microbiology and Biotechnology, Obolensk, Moscow Region, Russia, 142279

**Tatiana I. Kombarova** State Research Center for Applied Microbiology and Biotechnology, Obolensk, Moscow Region, Russia, 142279

**Nino Kupatadze** Institute of Chemistry and Molecular Engineering, Agricultural University of Georgia, 13 km. David Agmashenebeli Alley, 0131 Tbilisi, Georgia

**Alain Le Coupanec** Laboratory of Neuroimmunovirology, INRS-Institut Armand-Frappier, Institut national de la recherche scientifique, Université du Québec, 531, Boulevard des Prairies, Laval, QC, Canada

**Ji-Yun Lee** Department of Microbial, Biochemical and Food Biotechnology, University of the Free State, P.O. Box 399, Bloemfontein 9300, South Africa

**Mathieu Meessen-Pinard** Laboratory of Neuroimmunovirology, INRS-Institut Armand-Frappier, Institut national de la recherche scientifique, Université du Québec, 531, Boulevard des Prairies, Laval, QC, Canada

**Tamar Memenishvili** Institute of Chemistry and Molecular Engineering, Agricultural University of Georgia, 13 km. David Agmashenebeli Alley, 0131 Tbilisi, Georgia

**Alexander N. Mokrievich** State Research Center for Applied Microbiology and Biotechnology, Obolensk, Moscow Region, Russia, 142279

**Muhammad G. Morshed** Faculty of Medicine, Department of Pathology and Laboratory Medicine, BC Public Health Microbiology and Reference Laboratory, University of British Columbia, Provincial Health Services Authority, 655 West 12th Avenue, Vancouver, BC V5Z 4R4, Canada

**Anindita Mukerjee** Department of Molecular Biology and Immunology, Graduate School of Biomedical Sciences, University of North Texas Health Science Center, Fort Worth, TX 76107, USA; Institute for Cancer Research, University of North Texas Health Science Center, Fort Worth, TX 76107, USA

**Anastasia Baranova** State Research Center for Applied Microbiology and Biotechnology, Obolensk, Moscow Region, Russia, 142279

**Bijaya Pant** Central Department of Botany, Tribhuvan University, Kirtipur, Kathmandu, Nepal

**Vitalii M. Pavlov** State Research Center for Applied Microbiology and Biotechnology, Obolensk, Moscow Region, Russia, 142279

**Carolina H. Pohl** Department of Microbial, Biochemical and Food Biotechnology, University of the Free State, P.O. Box 399, Bloemfontein 9300, South Africa

**Vijay Kumar Prajapati** Infectious Disease Research Laboratory, Department of Medicine, Institute of Medical Science, Banaras Hindu University, Varanasi, India; Department of Biochemistry, The Central University of Rajasthan, Kishangarh, Rajasthan, India

**Amalendu P. Ranjan** Department of Molecular Biology and Immunology, Graduate School of Biomedical Sciences, University of North Texas Health Science Center, Fort Worth, TX 76107, USA; Institute for Cancer Research, University of North Texas Health Science Center, Fort Worth, TX 76107, USA

**Xiao Shen** Ecole Polytechnique Federale de Lausanne, CH-1015 Lausanne, Switzerland

**Onkar Nath Shrivastav** Department of Physics, Banaras Hindu University, Varanasi, India

**Geoffrey L. Smith** Department of Pathology, University of Cambridge, Tennis Court Road, Cambridge CB2 1QP, UK

**Pavel Solov'ev** State Research Center for Applied Microbiology and Biotechnology, Obolensk, Moscow Region, Russia, 142279

**Shyam Sundar** Infectious Disease Research Laboratory, Department of Medicine, Institute of Medical Science, Banaras Hindu University, Varanasi, India

**Chantel W. Swart** Department of Microbial, Biochemical and Food Biotechnology, University of the Free State, P.O. Box 399, Bloemfontein 9300, South Africa

**Pierre J. Talbot** Laboratory of Neuroimmunovirology, INRS-Institut Armand-Frappier, Institut national de la recherche scientifique, Université du Québec, 531, Boulevard des Prairies, Laval, QC, Canada

**David Tugushi** Institute of Chemistry and Molecular Engineering, Agricultural University of Georgia, 13 km. David Agmashenebeli Alley, 0131 Tbilisi, Georgia

**Wouter van der Westhuizen** Department of Microbial, Biochemical and Food Biotechnology, University of the Free State, P.O. Box 399, Bloemfontein 9300, South Africa

**Jamboor K. Vishwanatha** Department of Molecular Biology and Immunology, Graduate School of Biomedical Sciences, University of North Texas Health Science Center, Fort Worth, TX 76107, USA; Institute for Cancer Research, University of North Texas Health Science Center, Fort Worth, TX 76107, USA

**Christine Wandrey** Ecole Polytechnique Federale de Lausanne, CH-1015 Lausanne, Switzerland

**Thakur Prasad Yadav** Department of Physics, Banaras Hindu University, Varanasi, India

**Nino Zavradashvili** Institute of Chemistry and Molecular Engineering, Agricultural University of Georgia, 13 km. David Agmashenebelii Alley, 0131 Tbilisi, Georgia

# Bacterial Resistance to Quaternary Ammonium Compounds (QAC) Disinfectants

**Robert Bragg, Arina Jansen, Marisa Coetzee, Wouter van der Westhuizen and Charlotte Boucher**

**Abstract** Control of bacterial diseases has, for many years, been dependent on the use of antibiotics. Due to the high levels of efficacy of antibiotics in the past other disease control options have, to a large extent, been neglected. Mankind is now facing an increasing problem with antibiotic resistance. In an effort to retain some antibiotics for human use, there are moves afoot to limit or even ban the use of antibiotics in animal production. The use of antibiotics as growth promoters have been banned in the European Union and the USA. The potential ban on the use of antibiotics to treat diseases in production animals creates a dilemma for man—suffer significant problem with bacterial infection or suffer from a severe shortage of food! There are other options for the control of bacterial diseases. These include vaccine development, bacteriophage therapy, and improved biosecurity. Vaccine development against bacterial pathogens, particularly opportunistic pathogens, is often very challenging, as in many cases the molecular basis of the virulence is not always clearly understood. This is particularly true for *Escherichia coli*. Biosecurity (disinfection) has been a highly neglected area in disease control. With the ever-increasing problems with antibiotic resistance—the focus should return to improvements in biosecurity. As with antibiotics, bacteria also have mechanisms for resistance to disinfectants. To ensure that we do not replace one set of problems (increasing antibiotic resistance) with another (increasing resistance to disinfectants) we need to fully understand the modes of action of disinfectants and how the bacteria develop resistance to these disinfectants. Molecular studies have been undertaken to relate the presence of QAC resistance genes in bacteria to their levels of sensitivity to different generations of QAC-based products. The mode of action of QAC on bacteria has been studied using NanoSAM technology, where it was revealed that the QAC causes disruption of the bacterial cell wall and leaking of the cytoplasm out of the cells. Our main focus is on the control of bacterial and viral

R. Bragg (✉) · A. Jansen · M. Coetzee · W. van der Westhuizen · C. Boucher
Department of Microbial, Biochemical and Food Biotechnology,
University of the Free State, PO Box 399, Bloemfontein 9300, South Africa
e-mail: braggrr@ufs.ac.za

R. Adhikari and S. Thapa (eds.), *Infectious Diseases and Nanomedicine II*,
Advances in Experimental Medicine and Biology 808,
DOI: 10.1007/978-81-322-1774-9_1, 

diseases in the poultry industry in a post-antibiotic era, but the principles remain similar for disease control in any veterinary field as well as in human medicine.

**Keywords** Disinfectant · Biosecurity · Antibiotic resistance · NanoSAM · Staphylococcus aureus

## Increasing Antibiotic Resistance and the Concerns for Human Health

One of the main subjects for discussion at the First International Conference on Infectious Diseases and Nanomedicine, which was held in Nepal in December 2012 was the ever-increasing problem of antibiotic resistance [1]. As the main focus of our research is on the control of poultry diseases, we have focused this review on the problems associated with antibiotic resistance in poultry.

Antibiotics have been used at subtherapeutic levels as growth promoters in animals almost since their discovery, but the problems associated with the development and spread of antibiotic resistance have been steadily increasing since the 1960s [2, 3]. This widespread and unrestricted use of antibiotics has led to a surge in antibiotic resistant bacterial-related cases such as the increase in *Salmonella* resistant to Fluoroquinolone reported in the United States of America, Asia, and Europe [4–6]. The development of resistance to antibiotics is usually associated with their overuse and abuse, as well as the acquisition of genetic elements encoded within plasmids [7–11]. These increases had been related to the use of Fluoroquinolone as growth promoter in animals [4, 6]. The use of Fluoroquinolone as growth promoters in animal production was subsequently banned to reduce the incidence of bacterial resistance to Fluoroquinolone, but soon after, the use of therapeutic antibiotics increased to prevent and control bacterial infections in animals [12]. More restrictions on the use of antibiotics have been imposed and this led to an increase in the search for possible alternatives to control bacterial diseases in animal production [13]. Such alternative treatment methods include bacteriocins, small antimicrobial peptides, and bacteriophages [13]. These methods seem promising but they are still in the developmental stage and the safety applications of these methods are uncertain.

## Virulence in Avian Pathogenic *Escherichia coli* and the Difficulties in Producing an Effective Vaccine

One of the most important pathogens in poultry is *E. coli*. This bacterium is currently controlled through the widespread use of antibiotics in the poultry industry. Virulent strains of *E. coli* can obtain virulence genes through horizontal gene transfer and prophages from lysogenic bacteriophage infections [14]. Pathogenic *E. coli* are classified by their virulence properties such as ETEC (Enterotoxigenic *E. coli*), EPEC (Enteropathogenic *E. coli*), and EHEC (Enterohemorrhagic *E. coli*). Avian Pathogenic *E. coli* (APEC), the causative agent of the disease colibacillosis are *E. coli* strains found to be virulent to poultry [15].

APEC strains associated with colibacillosis may contain a variety of genes which may act individually or polygenically [16, 17]. The accurate identification of virulent strains of *E. coli* is essential for the potential development of any effective vaccine against colibacillosis in poultry.

In a study by Van der Westhuizen and Bragg [18], 18 different APEC virulence genes were characterized in South African *E. coli* isolates. The virulence genes were selected either from the literature which had shown high frequencies of these genes in APEC in other countries or from investigation on the contribution of these genes to virulence [19–22].

Virulence genes associated with APEC include the frz operon which promotes growth in serum during oxygen-restricted conditions [23], vacuolating autotransporter toxin gene which induces cytotoxic effects in the host cells [24], the *sitA* and *sitD* genes which are part of the sitABCD system, which is classified as a bacterial iron transporter [25], capsule formation transporter *kpsM* gene which codes for a protein transporter of polysaccharides for the formation of protective capsules [26]. Type 1 fimbrial adhesion gene which contributes to protection from host heterophils [27] is also one of the virulence genes of APEC. A novel APEC fimbrial adhesin gene which contributes to the colonization of the avian lung tissue [28] is another of the characterized virulence genes. The salmochelin siderophore receptor gene which facilitates iron chelation in the host [29] and the gene encoding for the episomal outer membrane protease which cleaves colicins [30] also play a role in virulence. A putative avian hemolysin gene [31] and an aerobactin siderophore receptor gene which contributes to iron uptake by the bacterium [32] have also been associated with virulence. The episomal increased serum survival gene *iss*, which aids in the resistance to host serum [33] and the pstSCAB system which has been shown to increase resistance to polymyxin, rabbit serum and acid shock [34] aid the bacterium in overcoming the host defence mechanisms and thus also play a role in virulence. The ColBM plasmid which contains a putative virulence cluster including colicin B and M production genes [35] and a transcriptional regulator of iron uptake genes in APEC, *uvrY* [36] have been identified in pathogenic *E. coli* strains isolated from poultry. The temperature-sensitive hemagglutinin *tsh* gene which facilitates hemagglutination activity in chickens [37]. The aerobactin iron acquisition operon iuc, which contributes to iron acquisition together with *sitA* and *FeoB* [25] play an important role in competing for iron in the host which results in increased virulence of the strains. Finally, a colicin V synthesis protein coded by the pAPEC-O2-ColV plasmid which was found to contribute to APEC virulence [38].

A vaccine that offers protection against *E. coli* infections in poultry has long been sought for and has not been possible until recent years. The main reason why a vaccine is not widely available has been due to the variation between APEC strains from different areas, as no one combination of different virulence genes seems to give APEC their virulence abilities. A recent study by Johnson et al. [21] included hundreds of different *E. coli* strains, including virulence profiles for each strain. This information was used to determine the most common occurring virulence genes among APEC strains and the five most prominent virulence genes were established.

Potential targets for vaccine production were now available but the localization of these virulence factors was also needed, to determine if these factors will encode proteins which will become useful epitopes on APEC cells. A study by Olsen et al. [39] provided these answers by theoretical calculations to determine the localization of various virulence proteins of APEC. The data from this study could then be combined with that of Johnson et al. [21] to provide promising targets for the development of a vaccine to prevent colibacillosis in poultry.

## Lack of Biosecurity and Good Disinfection

Biosecurity in the poultry industry is the least understood and least studied field of disease control. This also holds true for all other industries and for human health as well. Most poultry producers go through the process of disinfecting between cycles in the poultry sheds. Some even go to the lengths of monitoring the efficacy of the disinfection process, by taking samples for culturing of bacteria. However, the lack of understanding of the principles of good biosecurity often result in poor or inadequate disinfection, which leads to an increased number of mortalities in the flocks. This also holds true in the hospitals where confusion between nursing staff and domestic personnel over the allocation of disinfecting responsibilities often results in poor disinfection [40].

We have come a long way from the days where Joseph Lister [41] (1867) was the first to suggest that good biosecurity in a hospital is essential for control of bacterial infections. Surgeons no longer take pride in the blood stains on unwashed operating gowns and washing of hands is now mandatory. The question is, has much improved since 1867? Unfortunately, the answer to this question must be "No." Nosocomial infections currently are the fourth leading cause of disease in developed countries [42]. An estimated 1.75 million cases of hospital acquired infection per year occur in Europe [42], while Ecker and Carroll [43] estimated 2 million cases of hospital acquired infections in the USA each year with around 90,000 ending in the death of the patients. With these staggering numbers in developed countries, the question of the number of deaths from hospital acquired infections in developing countries would be substantially higher. With the ever-increasing problems of resistance to antibiotics, the numbers of hospital acquired infections are on the increase [44].

Lister [41] stated that hand washing is essential for disease control, yet this simple task still appears to be a problem, even in intensive care units [45]. This highlights the very severe problem of a lack of understanding of the importance of disinfection and biosecurity among healthcare workers and also suggests that little has changed from 1867 to 2013. Very few studies have investigated the risk for different patient categories and the disinfection and biosecurity regiments [40].

With the ever-increasing problems with antibiotic resistance in bacterium, it is high time that disinfection and biosecurity in the hospital environment received a much needed overhaul.

Many poultry producers accept that they will lose a percentage of the chickens in each batch and this is regarded as normal and a consequence of growing chickens. Similarly, healthcare works also accept that hospital acquired infections is a fact of life and nothing can be done about this. The development of a Virukill® continual disinfection program for use in poultry has demonstrated that good biosecurity practises can significantly reduce the clinical signs of the bacterial disease, infectious coryza in poultry [46]. This continual disinfection program was also shown to control Newcastle disease, a devastating viral infection of poultry, which can result in 100 % mortality [47]. Apart from disease control by good biosecurity in the face of disease challenge, it was also demonstrated that the correct use of disinfection and good biosecurity could reduce the "normal" mortality rates in poultry production by as much as between 20 and 50 % [47]. If good biosecurity and disinfection can reduce the severity of bacterial and viral diseases in poultry, these principles should also be applied to the hospital situation in an effort to reduce the unacceptably high numbers of hospital acquired bacterial infections.

## Resistance in Bacteria to Disinfectants

Microorganisms differ in their response to antimicrobial agents due to differences in their cellular structures, composition, and physiology [48]. In the literature, two major mechanisms of resistance to disinfectants are described, namely intrinsic and acquired resistance [49]. Intrinsic resistance is the natural ability of the bacterial cells to have decreased sensitivity to a specific agent. Intrinsic resistance is mostly demonstrated by Gram-negative bacteria [50]. The second mechanism is named acquired resistance which result from mutations of normal cellular genes, or the acquisition of foreign genetic information that allows the microorganism to survive harsh environments.

A disinfectant is only effective if it can reach and interact with its bacterial target site(s). Therefore, a disinfectant must cross the outer layers of the cell. The nature and composition of the outer layers may act as a permeability barrier, reducing the uptake of the disinfectant [48]. Gram-positive bacteria seem to be generally more susceptible to biocide action than gram-negative bacteria due to the lack of an outer membrane [51]. Another form of intrinsic resistance is the formation of biofilms by some bacterial cells, resulting in reduced sensitivity toward the disinfectant [48]. One possibility of the decrease in sensitivity could be based on a chemical interaction between the disinfectant and the biofilm, modulating of the microenvironment, production of degradative enzymes and neutralizing chemicals, or genetic exchange between cells in a biofilm [48].

Acquired resistance can arise by either mutation or by the acquisition of genetic material in the form of plasmids or transposons [48]. When microbial communities are exposed to either subinhibitory concentrations or long-term exposure to QACs with low-chemical reactivity, more resistant clones can emerge with changes in their susceptibility [50]. This can also cause changes in susceptibility to other antimicrobial agents [50, 52]. Although antimicrobial resistance can be obtained through mutations in the genes, it is proposed that antimicrobial resistance is mainly through the acquisition of antimicrobial resistance genes [50].

Acquired resistance against QACs is particularly seen in *Staphylococcus aureus* [53]. Several QAC resistance genes have already been identified from bacteria isolated from various sources. It was found that these genes are generally plasmid-borne. They encode proteins responsible for the expulsion of hydrophobic drugs including QACs, intercalating dyes and other cationic biocides. These genes are found on mobile genetic elements and their location allows them to interact between various staphylococci species [50].

The QAC resistance genes termed *qacA* and *qacB* are found on large plasmids, whereas the genes termed *smr*, *qacG*, *qacH*, and *qacJ* are found on plasmids smaller than 3 kb. The resistance genes found on small plasmids are commonly found in gene cassettes, and encode the protein part of the small multidrug resistance (SMR) family [54]. The *qacA* and *qacB* genes encode the protein part of the major facilitator superfamily (MFS) [55]. All the proteins encoded by the resistance genes are membrane embedded [56].

Efflux is the primary resistance mechanism of disinfectant resistance [57]. This lowers the concentration of the antimicrobial agent inside the cell by pumping the agent out of the cell [58]. Efflux pumps can either be substrate specific or have a wide range of substrates, such as the multidrug resistance pumps [58]. The multidrug resistance (MDR) proteins confer resistance to a wide range of substrates including toxic compounds, antibiotics, and QACs [52]. The antimicrobials are actively expelled out of the bacterial cell by these proteins [52]. These efflux systems are capable to pump out a broad range of chemically and structurally unrelated compounds in an energy-dependant manner, without altering the compound [59].

There are currently five families of multidrug efflux transporters recognized, namely: (1) The adenosine triphosphate (ATP)-binding cassette (ABC) transporters; (2) The small multidrug resistance (SMR) family; (3) The major facilitator superfamily (MFS); (4) The resistance-nodulation-division (RND) family, and (5) The multidrug and toxic compounds extrusion (MATE) family [52, 59]. The ABC transporters are the only family that belongs to the class primary transporters and the remaining four families belong to class secondary transporters. The MFS, SMR, and RND families are $H^+$/drug antiporters, whereas MATE is an $Na^+$/drug antiporter. These families of multidrug transporters also include proteins involved in PMF-dependent transport processes or other functions [60].

In the literature, it is proposed that there is a link between antibiotic and disinfectant resistance [50]. It was shown that some efflux pumps mediate cross-resistance since they export both QACs and other antimicrobial agents [50]. Co-resistance also exists, due to the linkage of different resistance mechanisms on the same genetic unit such as the same plasmid, transposon or integron, or a combination of both. Some efflux systems are able to accommodate both QACs and antimicrobials, due to the resistance determinants being on the same genetic unit. The QACs is thus able to select for organisms expressing efflux systems. Therefore, resistance can also be toward antibiotics [50, 61].

Disinfectants have been used longer in clinical environments than antibiotics. Thus, the *qac* gene is normally found on class 1 integrons, which is widespread in clinical isolates. The class 1 integrons found in environmental bacteria contains diverse *qac* gene cassettes [62]. The clinical class 1 integron might therefore have originated from a chromosomal class 1 integron, which was mobilized into a transposon. The initial selective advantage was conferred by resistance to disinfectants, mediated by the *qac* gene. Therefore, it seems that the increasing use of QACs promote the fixation of other novel genetic elements [50, 62].

The presence of QAC and other resistance determinants on the same mobile elements may contribute to transfer of resistance to other bacteria. The integrons are collectors of resistance cassettes and can readily pick up new additional resistance determinants on plasmids. This is an efficient vehicle system to travel among different species.

## Resistance to Disinfectants in *Staphylococcus aureus*

During the study, four *S. aureus* resistance strains were obtained. *S. aureus* strain ATCC 25923 was obtained to be a representative of a strain susceptible to QAC, thus the control strain. The minimal inhibitory concentration (MIC) for the QACs, benzalkonium chloride (BC), alkyldimethylammonium chloride (AAC), didecyl-dimethylammonium chloride (DDAC), and Virukill against the selected strains were tested. Virukill was selected as a representative of a commercially available QAC and DDAC because it is the active ingredient in Virukill, BC was selected as a representative of a first-generation QAC and AAC as it is an active ingredient in many QACs. The QAC genes, *smr, qacG, qacJ*, and *qacH* were identified in the respective resistant strains.

Only one gene was amplified and identified in the strains VB1_qacG, VB2_qacH, VB3_qacJ VB4_smr, and VB5_qacA but two genes were amplified and identified in the strains VB6 and APEC. None of the genes could be amplified from the control strain, ATCC 25923.

The minimal inhibitory concentration (MIC) against the strains containing the QAC-resistant bacteria confirmed the findings by [63] that the presence of these genes confer resistance to QACs in these organisms. The control strain ATCC 25923 was resistant to BC even though none of the *qac* genes could be amplified.

All of the strains tested, irrespective of the confirmed presence of a QAC resistance gene, were susceptible to Virukill.

The MIC recorded were significantly lower than the concentration generally used in disinfection, but the possibility still remains that the organisms might be exposed to concentrations low enough for resistant clones to survive. The MICs of these bacteria could potentially be much higher in a setting outside the laboratory as they could grow in conditions such as biofilms that would make them much more resistant to biocides [64].

The presence of the *qac* resistance genes in strains isolated from poultry is very important, because this is an area where disinfectants and antibiotics are frequently used and organisms may come into contact with both, thus a link between antibiotic and disinfectant resistance is possible.

Using real-time PCR the *qac* resistance genes used were amplified in most of the resistant strains including amplification of the *qac* resistance genes from the control strain, ATCC 25923. This indicates that more than one SMR efflux genes could be present in the resistant strains. This opens the door to more questions as to why only certain genes were amplified with conventional PCR and this prompted the investigation into the expression of these genes when induced with QACs.

A second hypothesis was formulated and this was that the *qac* genes were induced in the presence of QAC and thus the expression should be increased in the presence of a QAC. In order to perform expression studies, real-time PCR was used again as this is the technique mostly used for expression studies today because of its sensitivity and speed.

Induction of the *smr* gene was investigated in the *S. aureus* strains VB3_qacJ, VB4-smr, and ATCC 25923. Strain VB4-smr was used as a positive control as this strain was the only strain in which the *smr* gene was previously amplified using conventional PCR. Strain VB3-qacJ was chosen as a representative of a QAC-resistant bacterial strain and this strain was the only strain in which the *qacJ* gene was amplified using conventional PCR and the *smr* gene was amplified in this strain with real-time PCR from DNA extracted from this strain. The *smr* gene was present in high concentration in the strain VB4_smr during control conditions, i.e., conditions where no DDAC was present and not in the strains ATCC and VB3_qacJ. The study revealed that with the addition of DDAC there was no significant increase in the induction of the *smr* gene in strainVB4_smr and no induction was recorded for the strains VB3_qacJ or ATCC 25923. While looking at the expression we found a tenfold decrease in the expression of the *smr* gene in the case of DDAC treatment compared no DDAC treatment. There might be other genes and mechanisms involved in the process. We are currently looking at the expression of different genes during growth in DDAC.

Didecyldimethylammonium chloride (DDAC) is a quaternary ammonium compound (QAC) disinfectant often used in the poultry industry to disinfect hard surfaces. *Staphylococcus aureus* strain ATCC25923 treaded with DDAC revealed protuberances or bleb formations on their cell walls when observed with scanning electron microscopy. The DDAC-treated cells were further investigated using NanoSAM. The NanoSAM technology has been reviewed in this issue [65]. This

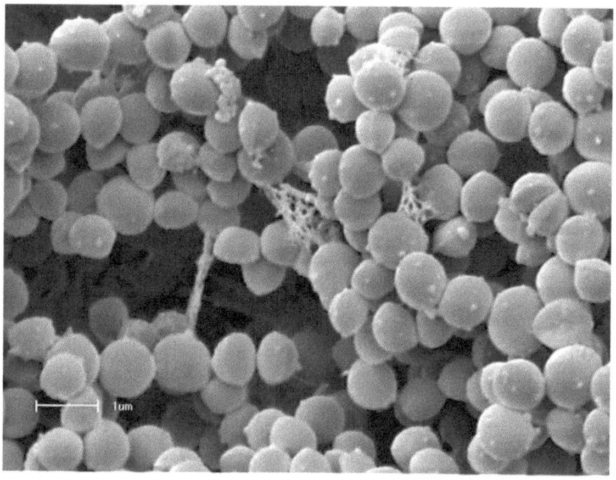


**Fig. 1** Electron micrograph of *Staphylococcus aureus* cells treated with QAC. Bleb formation (*BL*) and cellular material leakage (*CL*) can be observed

**Fig. 2** NanoSAM micrograph of *Staphylococcus aureus* cells with bleb formation

technology showed morphological changes as well as structural detail on control cells caused by the disinfectant that scanning electron microscopy could not (see Fig. 1). Elemental analysis was performed by focusing the nanoprobe on a specific target site on the sample which was bombarded with electrons resulting in the release of Auger electrons giving a specific energy profile [66]. These profiles were specific for every element and due to all elements having a unique set of electron-binding energies, the unknown elements in the sample could be identified [67]. SAM was achieved by mapping the various elements across the etched surfaces in the sample indicating the different elements in specific colors. NanoSAM showed a decrease in the elemental intensities during the etching of the cells treated with QAC. This proved that QAC leads to the leakage of cellular material (see Fig. 2). There was no chlorine detected in any of the samples which raised the question as to what happened to the QAC after the cells were exposed to it.

## Conclusions

With the ever-increasing concerns related to antibiotic resistance, there is a need to investigate alternatives to antibiotics for the control of bacterial diseases. One option is to investigate the use of bacteriophages as a potential therapy. This aspect has been reviewed in this issue [68] by our research team. Another possibility is to improve on the development of vaccines. As a result of the wide variety of virulence factors, the effective use of a vaccine for the control of *E. coli* infections remains a challenge. The other option is to improve on biosecurity and disinfection. This aspect is least well understood of the disease control options not only in the poultry industry, but also in human medicine. The simple fact that there are astronomically high numbers of hospital acquired infecting in developed countries is a clear indication that biosecurity and disinfection is not up to the standard in the medical field.

As the use of disinfectants is probably the best option for disease control in a post-antibiotic era, every effort must be made to fully understand the mechanisms of resistance to disinfectants to ensure that man does not also misuse disinfectants to a point where the majority of bacterial species develop resistance to disinfectants as well as to antibiotics. Much of the research efforts of this research group have been focused on understanding the mode of action of QAC-based disinfectants and how bacteria can become resistant to these disinfectants.

## References

1. Bragg RR, Kock L (2013) Nanomedicine and infectious diseases. Expert Rev Anti-infect Ther 11:359–361
2. Feighner SD, Dashkevicz MP (1987) Subtherapeutic levels of antibiotics in poultry feeds and their effects on weight gain, feed efficiency, and bacterial cholyltaurine hydrolase activity. Appl Environ Microbiol 53:331–336
3. Gilbert P, Moore LE (2005) Cationic antiseptics: diversity of action under a common apithet. J Appl Microbiol 99:703–715
4. Herikstad H, Hayes PS, Hogan J, Floyd P, Snyder L, Angulo FJ (1997) Ceftriaxone-resistant *Salmonella* in the united states. J Paediatr Infect Dis 9:904–905
5. Wiuff C, Madsen M, Baggesen DL, Aarestrup FM (2000) Quinolone resistance among *Salmonella enterica* from cattle, broilers, and swine in Denmark. Microbiol Drug Resist 6:11–17
6. White DG, McDermott PF (2001) Biocides, drug resistance and microbial evolution. Curr Opinion Microbiol 4:313–317
7. Rao GO (1998) Risk factors for the spread of antibiotic-resistant bacteria. Drugs 55:323–330
8. Feinman SE (1999) Antibiotics in animal feeds – drug resistance revisited. ASM News 64:24–29
9. Georgala DL (1999) Report on microbial antibiotic resistance in relation to food safety. Advisory committee on the microbiological safety of food (ACMSF). HMSO, London
10. Magee JT, Pritchard EL, Fitzgerald KA, Dunstan FDJ, Howard AJ (1999) Antibiotic prescribing and antibiotic resistance in community practice: retrospective study, 1996–1998. B Med J 319:1239–1240

11. Dixon B (2000) Antibiotics as growth promoters: risks and alternatives. ASM News 66:264–265
12. Casewell M, Friis C, Marco E, McMullin P, Phillips I (2003) The European ban on growth-promoting antibiotics and emerging consequences for human and animal health. J Antimicrob Chemother 53:159–161
13. Joerger RD (2003) Alternative to antibiotics: bacteriocins, antimicrobial peptides and bacteriophages. Poul Sci 82:640–647
14. Boyd EF, Brüssow H (2002) Common themes among bacteriophage-encoded virulence factors and diversity among the bacteriophages involved. Trends Microbiol 10:521–529
15. Barnes HJ, Glisson JR, Fadly AM, McDougald LR, Swayne D (2003) Colibacillosis. In: Saif YM (ed) Diseases of Poultry, 11th edn. Iowa State University Press, Ames, p 631
16. Delicato ER, de Brito BG, Gaziri LC, Vidotto MC (2003) Virulence-associated genes in *Escherichia coli* isolates from poultry with colibacillosis. Vet Microbiol 94:97–103
17. Vandekerchove D, Vandemaele F, Adriaensen C, Zaleska M, Hernalsteens JP, De Baets L et al (2005) Virulence-associated traits in avian *Escherichia coli*: comparison between isolates from colibacillosis-affected and clinically healthy layer flocks. Vet Microbiol 108:75–87
18. Van der Westhuizen WA, Bragg RR (2012) Multiplex polymerase chain reaction for screening avian pathogenic *Escherichia coli* for virulence genes. Avian Path 41:33–40
19. Rodriguez-Siek KE, Giddings CW, Doetkott C, Johnson TJ, Nolan LK (2005) Characterizing the APEC pathotype. Vet Res 36:241–256
20. Ewers C, Li G, Wilking H, KieBling S, Alt K, Antào E et al (2007) Avian pathogenic, uropathogenic, and newborn meningitis-causing *Escherichia coli*: how closely related are they? Int J Med Microbiol 297:163–176
21. Johnson TJ, Wannemuehler Y, Doetkott C, Johnson SJ, Rosenberger SC, Nolan LK (2008) Identification of minimal predictors of avian pathogenic *Escherichia coli* virulence for use as a rapid diagnostic tool. J Clin Microbiol 46:3987–3996
22. Herren C, Mitra A, Palaniyandi SK, Coleman A, Elankumaran S, Mukhopadhyay S (2006) The barA-uvrY two-component system regulates virulence in avian pathogenic *Escherichia coli* O78:K80:H9. Infect Immun 74:4900–4909
23. Rouquet G, Porcheron G, Barra C, Répérant M, Chanteloup NK, Schouler C, Gilot P (2009) A metabolic operon in extraintestinal pathogenic *Escherichia coli* promotes fitness under stressful conditions and invasion of eukaryotic cells. J Bact 191:4427–4440
24. Parreira VR, Gyles CL (2003) A novel pathogenicity island integrated adjacent to the *thrW* tRNA gene of avian pathogenic *Escherichia coli* encodes a vacuolating autotransporter toxin. Infect Immun 71:5087–5096
25. Runyen-Janecky LJ, Reeves SA, Gonzales EG, Payne SM (2003) Contribution of the *Shigella flexneri* sit, iuc and feo iron acquisition systems to iron acquisition in vitro and in cultured cells. Infect Immun 71:1919–1928
26. Pavelka MS Jr, Wright LF, Silver RP (1991) Identification of two genes, *kpsM* and *kpsT*, in region 3 of the polysialic acid gene cluster of *Escherichia coli* K1. J Bact 173:4603–4610
27. Mellata M, Dho-Moulin M, Dozois CM, Curtiss R III, Brown PK, Arnè P, Brèe A, Desautels C, Fairbrother JM (2003) Role of virulence factors in resistance of avian pathogenic *Escherichia coli* to serum and in pathogenicity. Infect Immun 7:536–540
28. Antão E, Ewers C, Gürlebeck D, Preisinger R, Homeier T, Li G, Wieler LH (2009) Signature-tagged mutagenesis in a chicken infection model leads to the identification of a novel avian pathogenic *Escherichia coli* fimbrial adhesin. PLoS One 4:e7796
29. Bäumler AJ, Norris TL, Lasco T, Voigt W, Reissbrodt R, Rabsch W, Heffron F (1998) IroN, a novel outer membrane siderophore receptor characteristic of *Salmonella enterica*. J Bact 180:1446–1453
30. Cavard D, Lazdunski C (1990) Colicin cleavage by ompT protease during both entry into and release from *Escherichia coli* cells. J Bact 172:648–652

31. Morales C, Lee MD, Hofacre C, Maurer JJ (2004) Detection of a novel virulence gene and a *Salmonella* virulence homologue among *Escherichia coli* isolated from broiler chickens. Foodborne Path Dis 1:160–165
32. Williams PH, Warner PJ (1980) ColV plasmid-mediated, colicin V-independent iron uptake system of invasive strains of *Escherichia coli*. Infect Immun 29:411–416
33. Chuba PJ, Leon MA, Banerkee A, Palchaudhuri S (1989) Cloning and DNA sequence of plasmid determinant *iss*, coding for increased serum survival and surface exclusion, which has homology with lambda DNA. Mol Gen Gene 216:287–292
34. Lamarche MG, Dozois CM, Daigle F, Caza M, Curtiss R III, Dubreuil JD, Harel J (2005) Inactivation of pst system reduces the virulence of an avian pathogenic *Escherichia coli* O78 strain. Infect Immun 73:4138–4145
35. Johnson TJ, Johnson SJ, Nolan LK (2006) Complete DNA sequence of a ColBM plasmid from avian pathogenic *Escherichia coli* suggests that it evolved from closely related ColV virulence plasmids. J Bact 188:5975–5983
36. Li G, Ewers C, Laturnus C, Diehl I, Alt K, Dai J et al (2008) Characterization of a *yjjQ* mutant of avian pathogenic *Escherichia coli* (APEC). Microbiology 154:1082–1093
37. Provence DL, Curtiss R III (1994) Isolation and characterization of a gene involved in hemagglutination by an avian pathogenic *Escherichia coli* strain. Infect Immun 62:1369–1380
38. Skyberg JA, Johnson TJ, Johnson JR, Clabots C, Logue CM, Nolan LK (2006) Acquisition of avian pathogenic *Escherichia coli* plasmids by a commensal *E. coli* isolate enhances its abilities to kill chicken embryos, grow in human urine, and colonize the murine kidney. Infect Immun 74:6287–6292
39. Olsen RH, Christensen H, Bisgaard M (2012) Comparative genomics of multiple plasmids from APEC associated with clonal outbreaks demonstrates major similarities and identifies several potential vaccine-targets. Vet Microbiol 158:384–393
40. Dancer SJ (2011) Hospital cleaning in the 21st century. Eur J Clin Microbiol Infect Dis 30:1473–1481
41. Lister J (1867) On the antiseptic principle in the practice of surgery. Brit Med J 2(351):245–260
42. Guggenbichler JP, Assadian O, Boeswald M, Kamer A (2011) Incidence and clinical impact of nosocomal infections associated with implantable biomaterials – catheters, ventilator-associated pneunmonia, urinary tract infections. GMs Krankenhhyg Interdiszip 6 Doc 18. doi:10.3205/dgkh000175
43. Ecker DJ, Carroll KC (2005) Investments in high-payoff technologies could reduce toll of infections. ASM News 71:576–581
44. Lin YE, Stout JE, Yu VL (2011) Prevention of hospital acquired legionellosis. Curr Opin Infect Dis 24:350–356
45. Alsubaie S, bin Maither A, Alalmaei W, Al-Shammari AD, Tashkandi M, Somily AM, Alaska A, BinSaeed AA (2013) Determinants of hand hygien noncompliance in intensive care units. Am J Infect Control 41:131–135
46. Bragg RR (2004) Limitation of the spread and impact of infectious coryza through the use of a continuous disinfection programme. Onderstepoort J Vet Res 71:1–8
47. Bragg RR, Plumbtree P (2003) Continuous disinfection as a means to control infectious diseases in poultry. Evaluation of a continuous disinfection programme for broilers. Onderstepoort J Vet Res 70:219–229
48. McDonnell G, Russell AD (1999) Antiseptics and disinfectants: activity, action and resistance. Clin Microbiol Rev 12:147–179
49. Russell AD (1998) Bacterial resistance to disinfectants: present knowledge and future problems. J Hosp Infect 43:S57–S68
50. Hegstad K, Langsrud S, Lunestad BT, Scheie AA, Sunde M, Yazdankhah SP (2010) Does the wide use of quaternary ammonium compounds enhance the selection and spread of antimicrobial resistance and thus threaten our health? Microl Drug Resist 16:91–104

51. White DG, McDermott PF (2001) Biocides, drug resistance and microbial evolution. Curr Opin Microbiol 4:313–317
52. Buffet-Bataillon S, Tattevin P, Bonnaure-Mallet M, Jolivet-Gougeon A (2012) Emergence of resistance to antimicrobial agents: the role of quaternary ammonium compounds – a critical review. Int J Antimicrob Agents 39:381–389
53. Ioannou CJ, Hanlon GW, Denyer SP (2007) Action of disinfectant quaternary ammonium compounds against *Staphylococcus aureus*. Antimicrob Agents Chemother 51:296–306
54. Bjorland J, Steinum T, Kvitle B, Waage S, Sunde M, Heir E (2005) Widespread distribution of disinfectant resistance genes among Staphylococci of bovine and caprine origin in Norway. J Clin Microbiol 43:4363–4368
55. Heir E, Sundheim G, Holck AL (1999) Identification and characterization of quaternary ammonium compound resistant *Staphylococci* from the food industry. Int J Food Microbiol 48:211–219
56. Langsrud S, Sundheim G, Brogman-Stahsen R (2003) Intrinsic and acquired resistance to quaternary ammonium compounds in food-related *Pseudomonas* spp. J Appl Microbiol 95:874–882
57. Hassan K A, Baltzer S A, Paulsen I T, Brown M (2010) Pumping out biocides-cause for concern. Microbiol Aust November: 178–181
58. Piddock LJV (2006) Clinically relevant chromosomally encoded multidrug resistance efflux pumps in bacteria. Clin Microbiol Rev 19:382–402
59. Kumar A, Schweizer HP (2005) Bacterial resistance to antibiotics: active efflux and reduced uptake. Adv Drug Del Rev 57:1486–1513
60. Putman M, van Veen HW, Konings WN (2000) Molecular properties of bacterial multidrug transporters. Microbiol Mol Biol Rev 64:672–693
61. Carson RT, Larson E, Levy SB, Marshall BM, Aiello AE (2008) Use of antibacterial consumer products containing quaternary ammonium compounds and drug resistance in the community. J Antimicrob Chemother 62:1160–1162
62. Gillings MR, Xuejun D, Hardwick SA, Holley MP, Stokes HW (2009) Gene cassettes encoding resistance to quaternary ammonium compounds: a role in the origin of clinical class 1 integrons? ISME J 3:209–215
63. Paulsen IT, Brown MH, Sunstan SJ, Skurray RA (1995) Molecular characterization of the *Staphylococcal* multidrug resistance export protein QacC. J Bacteriol 177:2827–2833
64. Campanac C, Pineau 1, Payard A, Baziard-Mouysset G, Roques C (2002) Interactions between biocide cationic agents and bacterial biofilms. Antimicrob Agents Chemother 46:1469–1474
65. Swart CW, Pohl CH, Kock JLF (2013) Auger-Architectomics: introducing a new nanotechnology to infectious disease. In: Adhikari, Thapa (eds) Advances in Experimental Medicine, Biology. Infectious Diseases and Nanomedicine I, vol 807, Chapter 1
66. Calvo-Barrio L, Pérez-Rodríguez A, Alvarez-Garcia J, Romano- Rodríguez A, Barcones B, Morante JR, Siemer K, Luck I, Klenk R, Scheer R (2001) Combined in-depth scanning auger microscopy and raman scattering characterisation of CuInS2 polycrystalline films. Vacuum 63:315–321
67. Hochella MF, Harris DW, Turner AM (1986) Scanning auger microscopy as a high-resolution microprobe for geologic materials. Am Miner 71:1247–1257
68. Bragg RR, van der Westhuizen W, Lee J-Y, Coetsee E, Boucher CE (2013) Bacteriophages as potential treatment option for antibiotic resistant bacteria. In: Adhikari, Thapa (eds) Advances in Experimental Medicine, Biology. Infectious Diseases and Nanomedicine I, vol 807, Chapter 7

# Immunological Markers that Correlate with Protection Immunity Against Tularemia Infection

**Victoria V. Firstova, Alexander N. Mokrievich, Vitalii M. Pavlov, Aleksey A. Gorbatov, Tatiana I. Kombarova, Sergey F. Biketov and Ivan A. Dyatlov**

**Abstract** An efficient immune response to tularemia is dependent on a strong cell-mediated component. We tried to identify markers of cellular immune responses that indicate a vaccine efficacy against tularemia. BALB/c mice were immunized with mutant *F. tularensis* 15Δ23A and/or *F. tularensis* 15 NIIEG strains and then were challenged i.n. with *F. tularensis* Schu. We compared the influence of *F. tularensis* antigens (tularinum) in vitro on production of IL-1, IL-5, IL-6, IL-17, IFN-$\gamma$, and TNF-$\alpha$ by splenocytes obtained from intact mice and mice immunized with mutant *F. tularensis* 15Δ23A and/or *F. tularensis* 15 NIIEG strains. We also compared expression of CD28, CD154, TLR-2, and CD69 markers on CD4 and CD8 T-cells after activation with tularinum in vitro. We found that tularinum-induced $CD4^+$ T-cells increased TNF-$\alpha$ and IFN-$\gamma$ synthesis and expression of CD69 only in group mice with high degree of post immunization protection against *F. tularensis* Schu challenge. Estimation of CD69 expression on $CD3^+CD4^+$ cells and IFN-$\gamma$, TNF-$\alpha$ synthesis by $CD4^+$ T-lymphocytes could be useful for determination protect ability of antitularemia immunity.

**Keywords** Tularemia · Cell immunity · Lymphocyte · CD69 · IFN-$\gamma$ · TNF-$\alpha$

## Introduction

Protection against tularemia can be acquired by vaccination with live tularemia vaccine or by natural infection. Both natural infection and vaccination induce long-lasting humoral and cell-mediated specific immunity. Since *Francisella*

V. V. Firstova (✉) · A. N. Mokrievich · V. M. Pavlov · A. A. Gorbatov · T. I. Kombarova · S. F. Biketov · I. A. Dyatlov
State Research Center for Applied Microbiology and Biotechnology, Moscow, Obolensk, Russia 142279
e-mail: firstova@obolensk.org

R. Adhikari and S. Thapa (eds.), *Infectious Diseases and Nanomedicine II*, Advances in Experimental Medicine and Biology 808,
DOI: 10.1007/978-81-322-1774-9_2, 

grows intracellularly, an efficient immune response to tularemia is dependent on a strong cell-mediated component [1].

Cell-mediated immunity could be estimated by lymphocyte stimulation test or by delayed-type hypersensitivity test with allergen from *F. tularensis*. Traditional lymphocyte stimulation methods are laborious and time-consuming [2]. Skin test has a number of drawbacks, including problems with standardization, possible cross-reactions. A number of recent studies have elucidated cell-mediated immune responses as immune correlates of protection subsequent to various infections or after vaccination [3, 4]. Multifaceted descriptions of T-cell phenotypes show good correlation to protection against infectious pathogens. Based on several models of infectious diseases, it has been suggested that proliferation of polyfunctional $CD4^+$ or $CD8^+$ T cells and their production of IFN-$\gamma$, TNF-$\alpha$, and IL-2 are crucial parameters to define protection [5]. Both $CD4^+$ and $CD8^+$ T cells of tularemia-vaccinated individuals respond with proliferation to various protein antigens of *F. tularensis*, and the proliferative response is strictly associated with IFN-$\gamma$ production [6]. These studies reaffirmed the notion that cell-mediated immunity is a critical determinant of *F. tularensis* immunity. Determination of the expression of surface antigens is another method for evaluating the lymphocyte effectors functions. Because little is known about CD69, CD154, and CD28 antigen expressions on CD4 and CD8 T lymphocytes of intact and immunized mice after in vitro tularinum stimulation and correlation of phenotype changes with protection against tularemia, we designed a study to evaluate the expression of these markers on T lymphocytes using flow cytometry. All these molecules: CD69, CD154, and CD28, play an essential role in cell immune responses. The CD69 antigen has been identified as the earliest activation marker on the surfaces of antigen- or allergen-specific activated lymphocytes in vitro [7]. Once CD69 is expressed, it acts as a co-stimulatory molecule for T-cell activation and proliferation [8]. The CD28 receptor is stimulated during the contact of T cells with antigen-presenting cells. Therefore, CD28 receptor stimulation is required for T cell responses to antigens and for B cell responses to T-dependent antigens [9]. The results of many studies have indicated that CD40–CD154 interaction can upregulate co-stimulatory molecules, activate antigen-presenting cells (APCs), influence T-cell priming, and T-cell-mediated effector functions [10].

*The work purpose* was to find specific immunological markers that correlate with protection immunity to tularemia infection after immunization mice with avirulent strains of *F. tularensis*.

## Materials and Methods

### Bacteria Strains

*F. tularensis* 15 line NIIEG, *F. tularensis* 15Δ23A, *and F. tularensis* Schu were obtained from State collection of pathogen microorganism (SRCAMB, Obolensk, Russia). Bacteria were cultured on FT agar (SRCAMB, Obolensk) for 16 h at

37 °C. The bacteria were suspended in phosphate buffer solution (PBS) for immunization or challenge procedures. *F. tularensis* 15Δ23A is a derivate of vaccine strain *F. tularensis* subsp. *holarctica* 15 line NIIEG with deleted one copy of *iglC* gene and inactivated *recA* gene. *F. tularensis* Schu subsp. *nearctica* is a high virulent strain.

## Animal Experiments

BALB/c mice were used at the age of 9 weeks. All animals were divided into four groups (15 mice in each group). Mice were immunized subcutaneously (s.c.) once with *F. tularensis* 15Δ23A ($1 \times 10^3$ CFU); twice with *F. tularensis* 15Δ23A ($1 \times 10^3$ CFU), and after 4 weeks with *F. tularensis* 15Δ23A ($1 \times 10^4$ CFU) or *F. tularensis* 15 line NIIEG ($1 \times 10^4$ CFU). On 28th day after last immunization, 10 mice from each group were challenged intranasaly (i.n.) with *F. tularensis* Schu 100 DCL (1DCL is 30 CFU for BALb/c mice). Mice were monitored daily for survival. All control mice died 4–6 days after challenge. The other five mice of each group were investigated for cytokine activities and intracellular and extracellular markers on T-cells.

## Lymphocyte Stimulation

Tularinum was used as a stimulating agent. Tularinum (Microgen, Russia) is an allergen tularemical fluid for the skin application which contains $10^9$ heat-killed *F. tularensis* 15. On 28th day after last immunization, spleens of five mice from each group were isolated and splenocytes were obtained. Splenocytes were seeded at $1 \times 10^6$ cells/well in 100 μL culture medium per well in a 96-well plate. Cells were stimulated with tularinum ($10^8$ heat-killed *F. tularensis* 15/ml) in 100 μL or an equal volume of culture medium for 16–18 h at 37 °C with 5 % $CO_2$.

## Cytokine CBA

Concentrations of IL-1, IL-5, IL-6, IL-17, IFN-γ, and TNF-α were estimated. Cell culture supernatants were assayed using the BD™ Cytometric Bead Array (CBA) Mouse Th1/Th2 kit (BD Biosciences, Bender MedSystem, GmbH) according to the manufacturers' protocols. Analysis of sample data was performed with BD™ CBA Software (BD Biosciences, Bender MedSystem, GmbH).

### *Flow Cytometry Analysis of Surface Markers and Intracellular Cytokine Staining*

For intracellular cytokine staining after 16–18 h of recall stimulation, 5 μg/mL of Brefeldin A was added to the splenocytes cultures. Four hours later, plates were centrifuged for 3 min at 500x g and supernatants were removed. Cells were prepared for labeling with monoclonal antibodies against cell surface marker or intracellular cytokines as recommended by BD Bioscience. The following monoclonal antibodies conjugates were used: CD3-FITC, CD3-PerCP, CD4-APC, CD8PE, CD69-FITC, TLR-2-FITC, CD28, CD154-FITC, IFNg-PerCP-Cy5,5, and TNFα-PE (BD Biosciences, Bender MedSystem, GmbH). All samples were analyzed with FACSCalibur, BD cytometer. 15,000 events were acquired for each analysis. To identify cell subset of interest, we gated $CD3^{+}CD4^{+}$ and $CD3^{+}$ $CD8^{+}$ cell populations and then identified cell subsets positive for the functional markers CD69, TLR-2, CD28, CD154, IFNg, and TNFα.

### *Statistical Evaluation*

Differences between the means of groups were analyzed using Student's *t*-test and Fisher's exact test.

## Results

### *Immunization with* F. tularensis *15Δ23A and* F. tularensis *15 NIIEG and Protection Against Pulmonary Challenge with* F. tularensis *Schu*

To determine the efficacy of immunization with *F. tularensis* avirulent strains in conferring protective immunity, BALB/c mice were immunized with mutant *F. tularensis* 15Δ23A once or twice or first with *F. tularensis* 15Δ23A and 4 weeks later with *F. tularensis* 15 NIIEG strain and then were challenged i.n. with *F. tularensis* Schu. As shown in Table 1, mice s.c. immunized twice with *F. tularensis* 15Δ23A exhibited significant protection (100 % survival) against i.n. *F. tularensis* Schu challenge during the monitoring period of 1 month. Mice immunized with *F. tularensis* 15Δ23A and then with *F. tularensis* 15 line NIIEG were partly protected (60 % mice survival) against i.n. challenge.

**Table 1** Mice survival and cytokine concentration in supernatants of immunized mice splenocytes (spontaneous response and tularinum-induced)

| Immunization | Mice survival, % | Splenocyte in vitro activation | Cytokine concentrations, pkg/mL | | | | | |
|---|---|---|---|---|---|---|---|---|
| | | | IL-1 | IL-5 | IL-6 | IFN-$\gamma$ | TNF-$\alpha$ | IL-17 |
| PBS | 0 | Medium | n/d | n/d | n/d | n/d | n/d | n/d |
| | | Tularinum | n/d | n/d | 227,99 | 10,78 | 61,77 | 0 |
| *F. tularensis* 15Δ23, once | 0 | Medium | n/d | n/d | n/d | n/d | n/d | n/d |
| | | Tularinum | n/d | n/d | 183,54 | 4,06 | 106,71 | 53,95 |
| *F. tularensis* 15Δ23, twice | 100 | Medium | 36,32 | 73,67 | n/d | n/d | n/d | n/d |
| | | Tularinum | 40,98 | 89,88 | 57,42 | **299,17** | 41,37 | 120,87 |
| *F. tularensis* 15Δ23 and *F. tularensis* 15 line NIIEG | 60 | Medium | n/d | n/d | n/d | n/d | n/d | n/d |
| | | Tularinum | n/d | n/d | 172,52 | **55,67** | 78,57 | 158,04 |

*n/d* not detectable

## *Cytokine Secretion In Vitro*

We compare cytokines secretions of tularinum-induced and spontaneous splenocytes responses obtained from mice immunized with avirulent *F. tularensis* strains (Table 1). Concentrations of IL-1, IL-5, IL-6, IL-17, IFN-$\gamma$, and TNF-$\alpha$ in supernatants were measured by cytometry with CBA analysis. Cytokines concentrations of spontaneous response of splenocytes were below the detection tests limits probably because the samples were diluted in medium twice and levels of cytokines were low. IL-1 and IL-5 spontaneous production by splenocytes were detectable only in group of mice immunized with *F. tularensis* 15Δ23 twice: 36,32 pkg/ml and 74,67 pkg/ml, respectively.

We found that tularinum-induced splenocytes of intact mice increased production of IL-6, IFN-$\gamma$, and TNF-$\alpha$. Statistical significances between the antigen-induced responses of each group were estimated. The amount of IFN-$\gamma$ found in culture with tularinum was higher in splenocyte of mice immunized twice with *F. tularensis* 15Δ23A or with *F. tularensis* 15Δ23A and then with *F. tularensis* 15 line NIIEG than in supernatants of splenocytes of intact mice (299,17 and 55,67 vs. 10,78 pkg/ml). The amount of IL-17 found in cell cultures incubated with tularinum in all group mice immunized with avirulent *F. tularensis* strains was higher than in tularinum-induced splenocytes of intact mice. Tularinum-induced TNF-$\alpha$ and IL-6 secretion was detected in all splenocytes culture.

## *Lymphocyte Stimulation*

To determine specific subpopulation cellular sources of IFN-$\gamma$ and TNF-$\alpha$ in leukocytes we performed intracellular staining and flow cytometry analysis. Addition of tularinum to splenocyte cultures resulted in a significant increase in

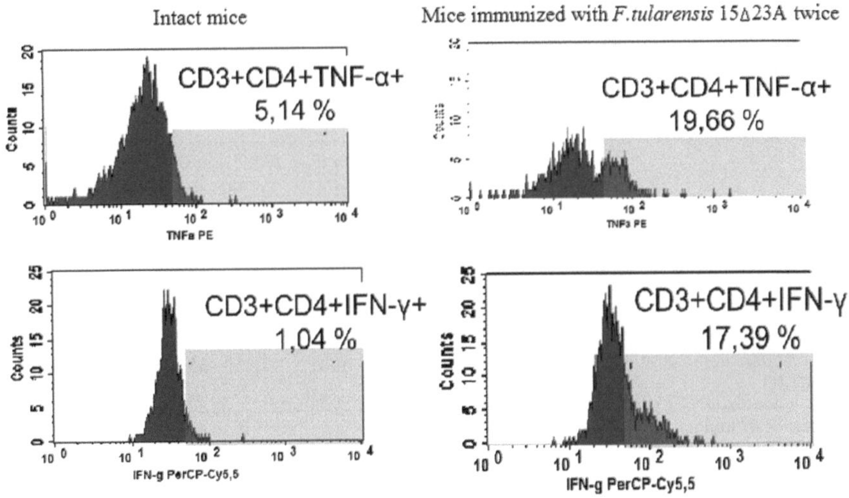


**Fig. 1** Production of TNF-α, IFN-γ by in vitro tularinum-stimulated $CD3^+CD4^+$ cells

IFN-γ, TNF-α producing T-cells (T-helper ($CD3^+CD4^+$) and cytotoxic ($CD3^+CD8^+$) lymphocytes) obtained only from mice groups, immunized twice with *F. tularensis* 15Δ23A or immunized with *F. tularensis* 15Δ23A and then with *F. tularensis* 15 line NIIEG (Fig. 1). There was no increase in IFN-γ, TNF-α production by splenocyte T-cell, obtained from mice groups immunized with *F. tularensis* 15Δ23A once.

We made comparative analyses of CD69, TLR-2, CD28, and CD154 expressions on T-cells after their stimulation with tularinum in vitro also. Addition of tularinum resulted in CD69 expression increase by splenocytes $CD4^+$T-cells obtained from mice groups, immunized twice with *F. tularensis* 15Δ23A or immunized with *F. tularensis* 15Δ23A and then with *F. tularensis* 15 line NIIEG (Fig. 2). The comparative analysis of TLR-2, CD28, and CD154 expressions on T-cells did not show remarkable differences among experimental mice groups.

## Discussion

*F. tularensis* 15 NIIEG is used as tularemia vaccinal prevention in Russia and showed good efficacy. However, *F. tularensis* 15 NIIEG cannot completely prevent pneumonic infection of type A bacteria. The exact mechanism of attenuation in *F. tularensis* 15 NIIEG is not elucidated yet. These problems can raise concern about vaccine safety. Development of tularemia vaccine with undoubted safety and rapid protection efficacy is actually still in process. Most scientists studying tularemia vaccine still keep trying to develop better mutant strains with higher immunogenicity and lower pathogenicity [11]. The efficacy of vaccine against

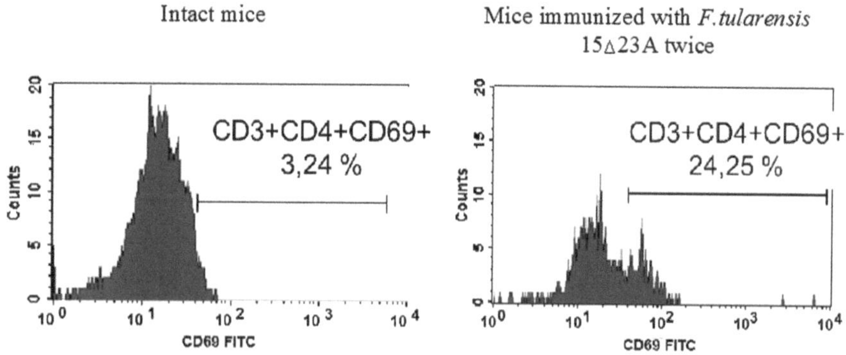


**Fig. 2** Expression of CD69 by CD3⁺CD4⁺ cells after their stimulation in vitro with tularinum

intracellular bacteria depends on the generation of cell-mediated immunity. We tried to identify cellular immune responses that indicate a vaccine efficacy against tularemia.

In our work we used a mutant of *F. tularensis* 15 NIIEG - *F. tularensis* 15Δ23A with deleted one copy of *iglC* gene and inactivated *recA* gene. We demonstrated that vaccination with *F. tularensis* 15Δ23A could protect 100 % of BALb/c mice against respiratory challenge with 3,000 CFU of *F. tularensis* Schu. Immunization with *F. tularensis* 15Δ23A and then with *F.tularensis* 15 line NIIEG also offered some protection against *F. tularensis* Schu strain in mice (60 % mice survival).

Similar to other intracellular pathogens, rapid production of proinflammatory and Th1-type cytokines is critical for initial control of *Francisella* infection [12]. It is well documented the importance of IFN-γ as an indicator of immunity against *F. tularensis* [13]. It is also known that responses to tularemia infection involve IL-17. Mechanistically, this is regulated via IL-17-mediated induction of IFN-γ and IL-12 in macrophages, linking Th17 and Th1 responses in vivo [14]. We analyzed concentrations of IL-1, IL-5, IL-6, IFN-γ, TNF-α, and IL-17 in supernatants of tularinum-induced and spontaneous splenocytes responses. It was found that IFN-γ and IL-17 were markedly increased after in vitro stimulation with tularinum in lymphocytes of mice that had been immunized twice with *F. tularensis* 15Δ23A (299,17 and 120,87 pkg/mL, accordingly) or with *F. tularensis* 15Δ23A and then with *F. tularensis* 15 line NIIEG (55,67 and 158,04 pkg/mL, accordingly) compared to tularinum-induced lymphocyte of intact mice (10,78 pkg/mL and not detectable concentration for IL-17, accordingly). We found that synthesis of IL-17 and TNF-α by splenocytes increased in response to tularinum recall in all immunized mice group. Intracellular staining showed that TNF-α and IFN-γ synthesis by CD4⁺ T-cells increased only in lymphocytes of immunized mice with protection against *F. tularensis* Schu i.n. challenge.

To evaluate the lymphocyte effector function, we determined the expression of surface antigens on CD4 and CD8 T-cells. TLR-2 is involved in the recognition of *F. tularensis* LVS by cells of mouse innate immune system. The role of individual

TLR in initiating essential responses to *F. tularensis* LVS may vary depending on the nature of host cell mediating protecting immune responses [15]. For these reasons, after stimulation with tularinum in vitro we compared TLR-2 expression on T-cells, obtained from immunized and intact mice. Mice immunization was not influenced on TLR-2 expression in tularinum-induced T-cells. Level TLR-2 in tularinum-induced T-cells (both CD4 and CD8) were increased in all mice group compare to noninduced T-cells.

CD69 is a type II membrane protein [16] expressed by T cells rapidly upon stimulation of the T-cell receptor (TCR) [17], which is why CD69 has been mostly regarded as an activation marker. Moreover, CD69 is required for the trafficking of effector CD4 T-cells to the bone marrow. Thus, CD69 is critical for the generation and maintenance of professional memory Th lymphocytes [18]. We found that CD69 expression increased in tularinum-induced $CD4^{+}$T-cells obtained from mice groups immunized twice with *F. tularensis* 15Δ23A or immunized with *F. tularensis* 15Δ23A and then with *F. tularensis* 15 line NIIEG, which had high protection against respiratory challenge with *F. tularensis* Schu (100 and 60 %, correspondingly).

We also analyzed co-stimulatory molecules CD28 and CD154. Expression of both CD28 and CD154 molecules increased on tularinum-induced T-cells of all mice groups.

## Conclusion

In agreement with the present results estimation of CD69 expression on $CD3^{+}CD4^{+}$ cells and IFN-$\gamma$, TNF-$\alpha$ synthesis by $CD4^{+}$T-lymphocytes could be useful for estimation of antitularemia immunity.

## References


1. Eneslätt K, Rietz C, Ryden P, Stöven S, House RV, Wolfraim LA, Tärnvik A, Sjöstedt A (2011) Persistence of cell-mediated immunity three decades after vaccination with the live vaccine strain of Francisella tularensis. Eur J Immunol 41:974–980
2. Syrjälä H, Herva E, Ilonen J, Saukkonen K, Salminen A (1984) A whole-blood lymphocyte stimulation test for the diagnosis of human tularemia. J Infect Dis 150:912–915
3. Millington KA, Innes JA, Hackforth S, Hinks TS, Deeks JJ et al (2007) Dynamic relationship between IFN-gamma and IL-2 profile of mycobacterium tuberculosis-specific T cells and antigen load. J Immunol 178:5217–5226
4. Winkler S, Necek M, Winkler H, Adegnika AA, Perkmann T et al (2005) Increased specific T cell cytokine responses in patients with active pulmonary tuberculosis from Central Africa. Microbes Infect 7:1161–1169
5. Seder RA, Darrah PA, Roederer M (2008) T-cell quality in memory and protection: implications for vaccine design. Nat Rev Immunol 8:247–258

6. Eliasson H, Per Olceñ, Sjöstedt A, Jurstrand M, Bäck E, Andersson S (2008) Kinetics of the Immune response associated with Tularemia: comparison of an enzyme-linked immunosorbent assay, a tube agglutination test, and a novel whole-blood lymphocyte stimulation test. Clin vaccine immunol 15(8):1238–1243
7. Testi R, D'Ambrosio D, De Maria R, Santoni A (1994) The CD69 receptor: a multipurpose cell-surface trigger for hematopoietic cells. Immunol Today 15:479–483
8. Ziegler SF, Ramsdell F, Alderson MR (1994) The activation antigen CD69. Stem Cells 12:456–465
9. Linsley PS, Ledbetter JA (1993) The role of CD28 receptor during T-cell responses to antigen. Annu Rev Immunol 11:191–212
10. Xu Y, Song G (2004) The role of CD40-CD154 interaction in cell immunoregulation. J Biomed Sci 11(4):426–438
11. Hong KJ, Park PG, Seo SH, Rhie GE, Hwang KJ (2013) Current status of vaccine development for tularemia preparedness. Clin Exp Vaccine Res 2(1):34–39
12. Sjöstedt A, Eriksson M, Sandström G, Tärnvik A (1992) Various membrane proteins of *Francisella tularensis* induce interferon-gamma production in both CD4 + and CD8 + T cells of primed humans. Immunology 76:584–592
13. Cowley S, Elkins K (2011) Immunity to Francisella. Front Microbiol 2:26
14. Onishi RM, Gaffen SL (2010) Interleukine-17 and its target genes: mechanisms of IL-17 function in disease. Immunology 129:311–321
15. Katz J, Zhang P, Martin M, Vogel SN, Michalek SM (2006) Toll-like receptor 2 is required for inflammatory responses to Francisella tularensis LVS. Infect Immun 74(5):2809–2816
16. Ziegler SF et al (1993) Molecular characterization of the early activation antigen CD69: a type II membrane glycoprotein related to a family of natural killer cell activation antigens. Eur J Immunol 23:1643–1648
17. Yamashita I, Nagata T, Tada T, Nakayama T (1993) CD69 cell surface expression identifies developing thymocytes which audition for T cell antigen receptor-mediated positive selection. Int Immunol 5:1139–1150
18. Shinoda K, Tokoyodaa K, Hanazawaa A, Hayashizakia K, Zehentmeier S, Hosokawaa H, Iwamura C, Koseki H, Tumesa DJ, Radbruch A, Nakayama T (2012) Type II membrane protein CD69 regulates the formation of resting T-helper memory Proc Natl Acad Sci USA 8. 109(19):7409–14

# Application of Plant Cell and Tissue Culture for the Production of Phytochemicals in Medicinal Plants

**Bijaya Pant**

**Abstract** Approximately 80 % of the world inhabitants depend on the medicinal plants in the form of traditional formulations for their primary health care system well as in the treatment of a number of diseases since the ancient time. Many commercially used drugs have come from the information of indigenous knowledge of plants and their folk uses. Linking of the indigenous knowledge of medicinal plants to modern research activities provides a new reliable approach, for the discovery of novel drugs much more effectively than with random collection. Increase in population and increasing demand of plant products along with illegal trade are causing depletion of medicinal plants and many are threatened in natural habitat. Plant tissue culture technique has proved potential alternative for the production of desirable bioactive components from plants, to produce the enough amounts of plant material that is needed and for the conservation of threatened species. Different plant tissue culture systems have been extensively studied to improve and enhance the production of plant chemicals in various medicinal plants.

**Keywords** Medicinal plants · Propagation · Conservation · Tissue culture

## Background

Medicinal plants have played a key role in health care system worldwide since the human civilization. Though total estimated flora of the world is not investigated yet, about 10 % of the plants all over the world are considered to be medicinal [1]. According to the estimation of the World Health Organization [2], approximately

B. Pant (✉)
Central Department of Botany, Tribhuvan University, Kirtipur, Kathmandu, Nepal
e-mail: pant_bijaya@yahoo.com; bijayapant@gmail.com

R. Adhikari and S. Thapa (eds.), *Infectious Diseases and Nanomedicine II*, Advances in Experimental Medicine and Biology 808,
DOI: 10.1007/978-81-322-1774-9_3, 

80 % of the world's rural people meet their needs of primary health care from traditional medicine [3]. Traditional medicinal systems such as Ayurvedic, Homeopathic, Unami, Naturopathy, Tibetan and even folklore medicines have served important role in the rural health care. This usually involves the use of plant extracts [4]. Many commercially used drugs have come from the information of indigenous knowledge of plants and their folk uses. Medicinal properties of plants are due to the active chemical constituents present in different parts of the plants, known as secondary metabolites, most of which probably evolved for chemical defenses against predation or infections. They are differentially distributed among limited taxonomic groups within the plant kingdom. Of the two types of metabolites primary and secondary produced by plants, the latter includes alkaloids, phenolics, flavonoids, steroids, tannins and terpenes [5], any modern drugs are derived from natural plants products or are chemical simulation of such substances. Most of the plants possess one or more of the medicinal properties, viz. antibacterial, antifungal, antiviral, antihelminthic, anticancer, laxative, sedative, cardio-tonic, diuretic and others [6]. Uses of medicinal and aromatic plants has greatly increased in recent years as herbal remedies, flavouring, pharmaceuticals, agrochemicals, pesticides food additives, perfumes, cosmetics and other natural products [7]. Demand of medicinal plants in these days is not only from Ayurvedic pharmaceutical companies but also from the modern pharmaceutical multinationals that are extensively investigating and exploring the possibilities of finding active pharmaceutical substances from these plants. Thus world market of medicinal plants and their product is increasing and their price is escalating due to their high demand. Himalaya is one of the richest regions of the medicinal and aromatic plants. Total number of species occurring in this region are not investigated properly, however it has been reported that over 10,000 species of medicinal plants occur in this regions [8]. The supply of the source plants, however, is often limited because of disease, changes in climate, changes in the economical development, their over exploitation is growing in the regions (Figs. 1, 2).

## Threat to Medicinal Plants

Increase in population and increasing demand of plant products along with illegal trade are causing depletion of plants species from their natural habitat making many valuable plants threatened and endangered. Most of these endangered species are the plants of medicinal and aromatic values, which are either confined in small areas and/or are collected recklessly from their natural habitat. The forest area in global context is decreasing in a fast pace which is leading to the extinction of many valuable medicinal herbs along with many other species not reported so far. Approximately 90 % of the medicinal plants that are used by the endogenous people are used for trade are collected from the wild. The plant material is harvested in the wild by gatherers, who usually collect whole plant. More than half of the plant material sold as traditional medicine consists of root, bulb, rhizomes or bark. When underground parts are harvested, the whole plant is removed, and too

**Fig. 1** Photograph of **a** wild plant of *Valeriana jatamansii,* a high value medicinal plant **b** micropropagated plants of *Valeriana jatamansii* (flowering) developed from nodal explants, and **c** a bunch of roots of *Valeriana jatamansii* for the production of essential oil in culture

often a tree is ring-barked, resulting in the death of that tree. It will be increasingly difficult to collect enough material from the wild, and several species can no longer be found. Threat to plant species has caused extinction of species, genetic erosion, and various agriculture related problems [9]. Thus, it is a universal concern for the

**Fig. 2** Photograph showing callus culture *Withania somnifera* (treated with colchicine) for the production of secondary metabolites

**Fig. 3** Photograph of *Dendrobium amoenum*, a beautiful medicinal orchid of Nepal (threatened species)

sustainable utilization of medicinal plants to support the livelihood of the people and conserve the threatened species for the conservation of biodiversity (Fig. 3).

Biotechnological tools especially in vitro culture technique has been investigated in various medicinal plants for their conservation and also as an alternative means of phytochemical production. Development of protocol for mass propagation of valuable medicinal plants has contributed not only in their conservation but also for their commercial cultivation.

## Plant Cell and Tissue Culture for Medicinal Plants

Plant cell and tissue culture offers an alternative source for rapid plant propagation and controlled production of disease-preventive phytochemicals or food ingredients in medicinal plants. Many plant species can be regenerated in vitro through

**Fig. 4** Photograph showing in vitro propagation through seeds of *Cymbidium aloifolium, a medicinal orchid* in MS medium, germination after 21 weeks of culture

several approaches such as from a single cell that can be reproduced, a tissue or organ part, or a cut out piece of differentiated tissue (or organ) known as an explant. These technologies can be established under sterile conditions from any vegetative parts or organs of plant such as meristems, nodes, leaves, stems, roots, axillary or adventitious shoot buds, endosperm, embryo etc. for unlimited plant multiplication and production of metabolites (Fig. 4).

## *Micropropagation in Medicinal Plants*

Micropropagation or clonal propagation is the vegetative propagation of plants in vitro in which multiplication of genetically identical copies of cultivars is brought about by asexual reproduction. Micropropagation offers the potential to produce thousands, or even billions of plants per year. Now a days, this technique is used for the clone selection, rapid and mass production of disease free materials of desired species in various organization or institutes for mass scale production of superior plants.

Micropropagation protocols have been developed for many important species of medicinal plants such as *Swertia chirayita*, *Cathranthus roseus*, *Panax, Stevia rebaudiana*, *Artemisia annua-aremisin*, *Elettaria cardamomum*, *Allium chinense, Camellia sinensis*, and so on. Many important medicinal herbs have been successfully propagated in vitro since long before either by organogenesis [10–16] or by somatic embryogenesis [17–19]. In vitro propagated plants of many important medicinal species were found to be uniform, showing less variation in their content of secondary metabolites than their wild/cultivated counterparts [20]. Micropropagation technique has also been developed in various plant species will allow for the improvement of phytopharmaceutical preparations thorough optimized techniques for the production of high-quality plant material (Fig. 5).

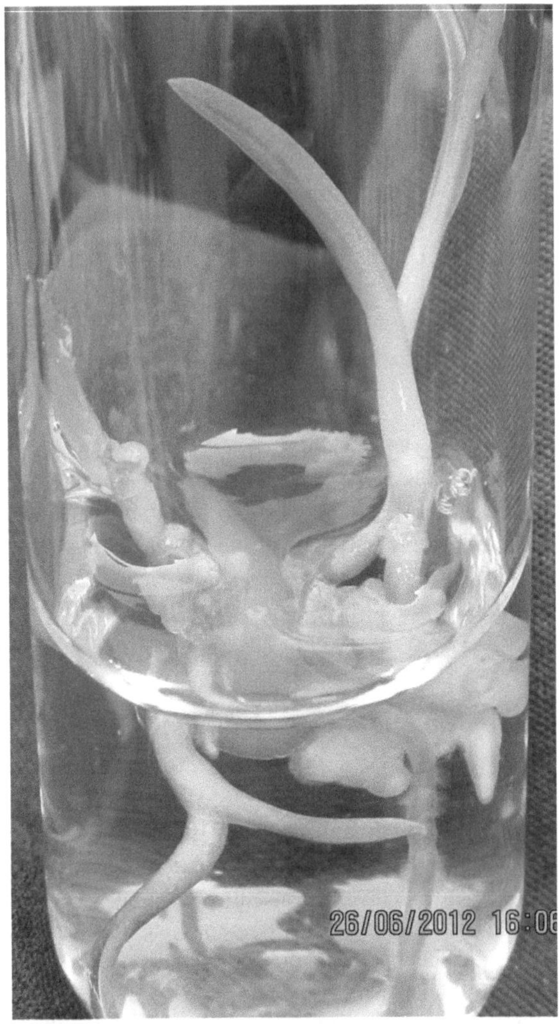


**Fig. 5** In vitro propagation of *Cymbidium aloifolium* through artificial seeds

## In Vitro Synthesis of Metabolites

Plant tissue cultures, are found to have potential as a supplement to traditional means of plant propagation in the industrial production of bioactive plant metabolites [21, 22]. Plant cells being totipotent; that is, cells in culture can produce the same metabolites as the whole plant. Many efforts have been made for commercial production of the medical plant metabolites from plant-cell culture. This is more convenient as useful compounds from the cells of any plants can be produced under controlled conditions independent of environmental conditions and these metabolites are more easily extractable from the culture cells than from the intact plants.

**Fig. 6** In vitro propagated plantlets of *Phaius tancarvilleae* (medicinal orchid) form shoot tip culture grown in an earthen vase

Conventional biotechnological methods for influencing secondary metabolite production include changing nutrient composition, environmental conditions, and growth regulator composition. More recently, use of specialized techniques such as metabolic engineering, cell immobilization, elicitation, and in situ product removal have been investigated and applied to increase yields of secondary metabolites. In vitro production of secondary metabolite in plant cell cultures has been reported from various medicinal plants. Various food ingredients and phytochemicals, including flavors, nutraceuticals, colorants, essential oils, sweeteners, antioxidants have been produced in cell culture [23] some of which have been commercialized [24, 25].

Advances in plant cell cultures could provide new means for the cost-effective, commercial production of even rare or exotic plants, their cells, and the chemicals that they will produce. A continuation and intensification efforts in this field will lead to controllable and successful biotechnological production of specific, valuable, and as yet unknown plant chemicals (Fig. 6).

## Types of Culture

Different strategies, using an in vitro culture system, have been extensively studied to improve the production of plant chemicals [26]. Several types of cell culture are used to produce the desired metabolites or food ingredients, such as organized tissue, callus, cell suspension, immobilized cells, transform cells and organs.

## *Organized Tissue Cultures*

Various organs such as shoot, root, and other plant organ cultures have been developed for the production of desired compounds that require cell differentiation. Though the production of secondary metabolites in organized tissue cultures differ from those of the intact plants, stable growth and consistent secondary metabolite production equivalent or in higher concentration have been observed in shoot and root cultures of many species [27–30]. In some cases, compounds not found in the intact plants have been detected in cultured tissue. The alkaloid levels of cultured roots of *Stephania cepharantha* were much higher than those of the plants and aromoline, which was not present in the original plant was found in cultured plants [31]. High yield production of tanshinones alkaloids by adventitious root cultures of *Salvia miltiorrhiza*, a well known traditional Chinese medicine in optimized culture conditions [32]. A protocol for complete plant regeneration via somatic embryogenesis from tuber derived callus of *Corydalis yanhusuo*, a rich source of several pharmacologically important alkaloids containing plants was standardized [33]. They have reported the production of higher concentration of bioactive compounds such as D, L-tetrahydropalmatine and D-corydaline from the tubers of somatic embryo-derived plants [33]. The content of gentiopicroside and swertiamarin in the aerial and underground parts of *Gentiana davidii* var. *formosana* was higher than in the marketed crude drug (underground parts of *G. scabra*) *davidii* var. *formosana* [34]. In *Fritillaria unibracteata* bulbs rapidly propagated from the small cuttings of the bulb explants produced higher content of alkaloid and beneficial microelements in the cultured bulbs than found in the wild bulb [35].

## Callus and Cell Suspension Culture

Callus is an undifferentiated mass of cell developed in culture from any kind of explants. Cell suspension cultures are rapidly dividing homogenous suspensions of cells grown in liquid nutrient media. In a cell suspension, a callus is suspended in liquid media. The cell suspension is then placed on a shaker to allow the cell aggregates to disperse to form smaller clumps and single cells that are equally distributed and continuously grow throughout the liquid media [36]. Thus plant suspension cultures provide a valuable platform for the production of high-value secondary metabolites and other substances of commercial interest [37]. Various workers have reported the production of secondary metabolites from callus culture [37–41]. Successful attempts to produce the valuable pharmaceuticals in relatively large quantities by callus and cell cultures have been achieved. Production of morphine and codeine in morphologically undifferentiated cultures of callus and suspension cultures has been reported in *Papaver somniferum* as an alternative means for the production of these compounds [42, 43]. Taxol is an anticancer drug approved for clinical treatment of ovarian and breast cancer by the Food and Drug Administration (FDA, USA). It also has significant activity against malignant melanoma, lung cancer, and other solid tumors [37, 44]. Production of taxol by various *Taxus* species cells in cultures has been one of the most extensively explored areas of plant cell cultures in recent years due to its enormous commercial value, scarcity of the *Taxus* tree, and the costly synthetic process [45]. The kinetics of biomass accumulation and paclitaxel production by *T. baccata* cell in suspension cultures was studied [46]. Paclitaxel was found to accumulate at high yields (1.5 mg/l) from *T. brevifolia* cell suspension culture [46, 47]. Two hundred mg/l taxol per liter of cell suspension culture were obtained in *Taxus mairei* by needle and stem explants derived callus on Gamborg's B5 medium [48]. Diosgenin, an important precursor for the chemical synthesis of steroidal drugs to the pharmaceutical industry has been obtained in cell suspension culture of *Dioscorea deltoidea* [49]. Similarly various other important metabolites such as camptothecin, a potent antitumor alkaloid were synthesized from callus culture of *Camptotheca acuminata* [50]. The root of *Panax ginseng* C.A. Mayer, so-called ginseng, has been widely used as a tonic and highly prized medicine since ancient times Ginseng has been recognized as a miraculous promoter of health and longevity. Cultured ginseng products derived from cell suspension culture of *P. ginseng* have been commercially produced by Nitto Denko Co. in Japan since 1990 [24, 51].

## Scale up of Secondary Metabolites by Using Different Systems

Advances in tissue culture, combined with improvement in genetic engineering, specifically transformation technology has opened new avenues for high volume production of pharmaceuticals, nutraceuticals, and other beneficial substances [52]. For the large-scale cultivation of plant cells for the production of valuable bioactive compounds specific bioreactors systems are applied. Plant cells cultured in liquid suspension must be accommodated in large-scale bioreactor for the constant and continuous production of metabolites. Various types of bioreactors have been designed and being used for the commercial production of metabolite [53]. Some of the culture systems to scale up of secondary metabolites are discussed here.

Cell immobilization is a technique to fix plant cells in a suitable matrix. The immobilization of plant cells protects the cells from mechanical stress and facilitating product recovery. In recent years the preparation and use of immobilized cells have drawn great interest in biotechnology and its range of applications is still expanding.Cell immobilization has been used to synthesize extracellular metabolites and for the biotransformation of many metabolites observing its positive effects on secondary metabolite production. Several food ingredients have also been produced by immobilized cell cultures. Some of the medicinal compounds localized in morphologically specialized tissues or organs of plants have been produced in culture systems not only by inducing specific organized cultures, but also by undifferentiated cell cultures. The possible use of plant cell cultures for the specific biotransformation of natural compounds has been demonstrated by various workers [54–57]. Capsaicin production was increased 1,000-fold after *Capsicum frutescens* cells were immobilized in the pores of a polyurethane foam matrix and cultured in Murashige and Skoog's basal medium containing 10 mM $CaCl_2$ [58]. For the production of plumbagin in Cell cultures of *Plumbago rosea* immobilized in calcium alginate enhanced the production of plumbagin an important medicinal compound [59].

## Elicitation

Elicitors are compounds that stimulate the synthesis of phytoalexins and various secondary metabolites in plants. Cell cultures are exposed to biotic elicitors such *Aspergillus niger*, crude chitin and abiotic elicitors such as mannitol, methyl jasmonate to induce metabolite production. Elicitors such as fungal wall materials, plant and microbial polysaccharides, and some chemicals increase secondary metabolite production in various plant cell and tissue culture systems. The most frequently used elicitors are fungal carbohydrates, yeast extract, MJ and chitosan. MJ, a proven signal compound, is the most effective elicitor of taxol production in *Taxus chinensis* and goinsenoside production in *P. ginseng* [60]. Anthraquinone production in *Rubia akane* cell culture was increased by chitosan, a biotic elicitor polysaccharide [61]. Several studies have shown that elicitation increased the production yield of many food ingredients and secondary metabolites in various types of plant cell and tissue culture [62–64]. Enhance rate of secondary metabolites production due to elicitation treatment is elicitation may activate novel genes that encode proteins with no similarities to those of known genes [65]. It is now known that elicitation ultimately affects the transcription rates of genes encoding enzymes involved in plant defense pathways [66].

## Transformed Tissue Cultures

Achievement of recombinant DNA technology has contributed a lot in the welfares of mankind introducing the technology beyond the imagination. *Agrobacterium rhizogenes*-mediated transformation has been used to produce transgenic hairy root cultures for secondary metabolite production [67, 68]. Transgenic hairy root cultures have revolutionized the role of plant tissue culture in secondary metabolite production. They are unique in their genetic and biosynthetic stability, faster in growth, and more easily maintained. Infection of wounded plants by *A. rhizogenes* bacteria results in hairy-root disease, which is characterized by rapidly growing and highly branched roots at the host wound sites. T-DNA of the bacterial

Ri (root-inducing) plasmid, which contains genes encoding endogenous hormones, is integrated into the plant genome and allows the proliferation of these adventitious roots. Similarly, transformation of plant tissue with T-DNA from *A. tumefaciens* leads to the formation of adventitious shoots or "shooty teratomas". Using this methodology a wide range of plant metabolites and food ingredients have been synthesized [69–71]. Hairy roots cultures grow faster than the untransformed roots or shoots, and in many cases, their growth rates approach those of cell suspension cultures with the stable genetic and biochemical characteristics [72]. The transformed roots have been found to produce the secondary metabolites equivalent or higher than that in the parent plants [73, 74]. Though there is variation in clone to clone in some transformation process various strategies are investigated for the stable transformation process [52].

## *Metabolic Engineering*

Metabolic engineering has been applied to improve production of secondary metabolites in different culture systems [75]. Metabolic engineering is the targeted and purposeful alteration of metabolic pathways in an organism to achieve better cellular pathways for chemical transformation and regulatory functions of the cells by using recombinant DNA technology [76]. Metabolic engineering directs cellular metabolism to produce desired metabolites by inserting genes that encode specific biosynthetic enzymes into the plant genome [57]. Efforts have been made for the production of specific metabolites such as scopolamine, nicotine, and berberine by altering the metabolic pathways [77].

# Conclusions

The existing knowledge of plant cell and tissue culture can be applied in medicinal plants using different strategies for the production and improvement of metabolites. It has been demonstrated that the biosynthetic activity of cultured cells can be enhanced by regulating environmental factors, as well as by artificial selection or the induction of variant clones or by gene transfer technology. Due to these advances, research in the area of tissue culture technology for production of plant chemicals has bloomed beyond expectations. As more is known about the biochemical and genetic regulation of plant secondary metabolism, and more advances are made in the development of yield improvement strategies and design of large-scale bioreactors, commercial development of plant cell and tissue culture derived metabolites and food ingredients is expected to increase in the coming years. Therefore the application of these technologies will allow for improved phytopharmaceutical manufacture and the in-depth investigation of medicinal plant physiology.

**Acknowledgments** All the tissue culture photographs are used from our research work. The author gratefully acknowledges all the team members involved in tissue culture research project.

## References

1. Schippmann U, Danna L, Cunningham AB (2006) A comparison of cultivation and wild collection of medicinal and aromatic plants under sustainability aspects. In: Bogers RJ, Craker LE, Lange D (eds) Medicinal and aromatic plants. Springer, New York
2. Bannerman RH (1982) Traditional medicine in modern health care. World Health Forum 3(1):8–13
3. Sarker KP, Islam A, Islam R, Hoque A, Joarder I (1996) In-vitro propagation of *Rauvolfia serpentina* L. Benth. ex Kurz. through tissue culture. Planta Med 62:358–359
4. Vieira RF, Skorupa LA (1993) Brazilian medicinal plants gene bank. Acta Hortic 30:51–58
5. Farnsworth NR (1966) Biological and phytochemical screening of plants. J Pharm Sci 55(3):225–276
6. Balandrin MJ, Klocke JA (1988) Medicinal, aromatic and industrial materials from plants. In: Bajaj YPS (ed) Biotechnology in agriculture and forestry, vol 4. Springer, Heidelberg, pp 1–36
7. Kumar S, Hassan SA, Dwived S, Kukreja AK, Sharma A, Singh AK, Sharma S, Tewari R (2000) Proceedings of the national seminar on the frontiers of research and development in medicinal plants. J Med Aromatic Plants Sci 22(4A):461–463
8. Senghi P (2001) Ethnobotanical approaches of traditional medicine studies: some experiences from Asia. Pharm Biol 39:74–79
9. Pant B, Swar S, Gurung R (2007) Current Status and *ex situ* conservation of threatened orchids of Nepal. In: Proceedings of 9th Asia Pacific Orchid Conference (APOC 9), Seol, Korea
10. Erdei I, Kiss Z, Maliga P (1981) Rapid clonal multiplication of *Digitalis lanata* in tissue culture. Plant Cell Rep 1:34–35
11. Nishioka I (1988) Clonal multiplication of medicinal plants by tissue culture. Shoyakugaku Zasshi 42:1–11
12. Tsay HS, Huang HL (1998) Somatic embryo formation and germination from immature embryo—derived suspension—cultured cells of *Angelica sinensis* (Oliv.) Diels. Plant Cell Rep 17:670–674
13. Pant B, Kohda H, Namera A (1996) Clonal propagation of *Cnidium officinale by* shoot tip culture. Planta Med 62:281–283
14. Nongdam P, Chongtham N (2011) In vitro rapid propagation of *Cymbidium aloifolium* (L.) Sw.: a medicinally important orchid via seed culture. J Biol Sci 11(3):254–260
15. Pant B, Thapa D (2012) In vitro mass propagation of an epiphytic orchid *Dendrobium primulinum* Lindl. through shoot tip culture. Afr J Biotechnol 11(42):9970–9974
16. Pradhan S, Paudel YP, Pant B (2013) Efficient regeneration of plants from shoot tip explants of *Dendrobium densiflorum* Lindl., a medicinal orchid. Afr J Biotechnol 12(12):1378–1383
17. Hiraoka N, Kodama T, Oyanagi M, Nakano S, Tomita Y, Yamada N, Iida O, Satake M (1986) Characteristics of *Bupleurum falcatum* plants propagated through somatic embryogenesis of callus cultures. Plant Cell Rep 5:319–321
18. Kitamura Y, Miura H, Sugii M (1989) Plant regeneration from callus cultures of *Swertia pseudochinensis*. Shoyakugaku Zasshi 43:256–258
19. Sagare AP, Lee YL, Lin TC, Chen CC, Tsay HS (2000) Cytokinin-induced somatic embryogenesis and plant regeneration in *Corydalis yanhusuo* (Fumariaceae) a medicinal plant. Plant Sci 160:139–147

20. Yamada Y, Shoyama Y, Nishioka I, Kohda H, Namera A, Okamoto T (1991) Clonal micropropagation of *Gentiana scabra* Bunge var. *buergeri* Maxim and examination of the homogeneity concerning the gentiopicroside content. Chem Pharm Bull 39:204–220
21. Ramachandra RS, Ravishankar GA (2002) Plant cell cultures: chemical factories of secondary metabolites. Biotechnol Adv 20:101–153
22. Dörnenburg H (2010) Cyclotide synthesis and supply: From plant to bioprocess. Biopolymers Pept Sci 94(5):602–610
23. Dörnenburg H, Knorr D (1996) Generation of colors and flavors in plant cell and tissue cultures. Crit Rev Plant Sci 15:141–168
24. Ushiyama K (1996) Commercial production of ginseng from plant cell cultures. Institute of Food Technologists Annual Meeting, New Orleans
25. Abdin MZ, Kamaluddin A (2006) Traditional systems of medicine: improving quality of medicinal herbs through physico-chemical and molecular approaches. India Publishing House Pvt. Ltd, New Delhi
26. Dörnenburg H, Knorr D (1997) Challenges and opportunities for metabolite production from plant cell and tissue cultures. Food Technol 51:47–54
27. Endo T, Yamada Y (1985) Alkaloid production in cultured roots of three species of *Duboisia*. Phytochemistry 24:1233–1236
28. Miura Y, Hirata K, Kurano N, Miyamoto K, Uchida K (1988) Formation of vinblastine in multiple shoot culture of *Cathranthus rosus*. Planta Med 54:18–20
29. Davioud E, Kan C, Hamon J, Tempe J, Husson HP (1989) Production of indole alkaloids by in vitro root cultures from *Catharanthus trichophyllus*. Phytochemistry 28:2675–2680
30. Li W, Li M, Yang DL, Xu R, Zhang Y (2009) Production of podophyllotaxin by root culture of *Podophyllum hexandrum* Royle. Electron J Biol 5:34–39
31. Sugimoto Y, Sugimura Y, Yamada Y (1988) Effects of culture conditions on roots of *Stephania cepharantha*. Agric Biol Chem 52:1495–1498
32. Shimomura K, Kitazawa T, Okamura N, Yagi A (1991) Tanshinone production in adventitious roots and regenerates of *Salvia miltiorrhiza*. J Nat Prod 54:1583–1586
33. Lee YL, Sagare AP, Lee CY, Feng HT, Ko YC, Shaw JF, Tsay HS (2001) Formation of protoberberine-type alkaloids by the tubers of somatic embryo-derived plants of *Corydalis yanhusuo*. Planta Med 67:839–842
34. Chueh FS, Chen CC, Sagare AP, Tsay HS (2000) Quantitative determination of secoiridoid glucoside in in vitro propagated plants of *Gentiana davidii* var. *formosana* by high performance liquid chromatography. Planta Med 67:70–73
35. Gao SL, Zhu DN, Cai ZH, Jiang Y, Xu DR (2004) Organ culture of a precious Chinese medicinal plant—*Fritillaria unibracteata*. Plant Cell Tiss Org Cult 59:197–201
36. King PJ (1984) Induction and maintenance of cell suspension cultures. In: Vasil IK (ed) Cell Culture and Somatic Cell Genetics of Plants. Academic Press, New York
37. Wickremesinhe ERM, Arteca RN (1993) *Taxus* callus cultures: initiation, growth optimization, characterization and taxol production. Plant Cell Tiss Org Cult 35:181–193
38. Baumert A, Groger D, Kuzovkina IN, Reisch J (1992) Secondary metabolites produced by callus cultures of various *Ruta* species. Plant Cell Tiss Org Cult 28:159–162
39. Thengane SR, Kulkarni DK, Shrikhande VA, Joshi SP, Sonawane KB, Krishnamurthy KV (2003) Influence of medium composition on callus induction and camptothecin(s) accumulation in *Nothapodytes foetida*. Plant Cell Tiss Org Cult 72:247–251
40. Shrestha SK, Pant B (2012) Production of Bergenin, an active chemical constituent in the callus of *Bergenia ciliata* (Haw.) Sternb. Bot orientalis 8:40–44
41. Paudel S, Adhikari SR, Pant B (2013) Effect of colchicine on production of secondary metabolites from callus of *Withania somnifera* (L.) Dunal. J Nepal Biotechnol Assoc 3(1):15–18
42. Yoshikawa T, Furuya T (1985) Morphinan alkaloid production by tissues differentiated from cultured cells of *Papaver somniferum*. Planta Med 2:110–113
43. Furuya T, Ikuta A, Syono K (1972) Alkaloids from callus cultures of *Papaver somniferum*. Phytochemistry 11:3041–3044

44. Wickremesinhe ERM, Arteca RN (1994) *Taxus* cell suspension cultures: optimizing growth and production of taxol. J Plant Physiol 144:183–188
45. Cragg GM, Schepartz SA, Suffuess M, Grever MR (1993) The taxol supply crisis: new NCI policies for handling the large-scale production of novel natural product anticancer and anti-HIV agents. J Nat Prod 56:1657–1668
46. Srinivasan V, Pestchanker L, Moser S, Hirasuma T, Taticek RA, Shuler ML (1995) Taxol production in bioreactors; kinetics of biomass accumulation, nutrient uptake, and taxol production by cell suspensions of *Taxus baccata*. Biotechnol Bioeng 47:666–676
47. Kim JH, Yun JH, Hwang YS, Byun SY, Kim DI (1995) Production of taxol and related taxanes in *Taxus brevifolia* cell cultures. Biotechnol Lett 17(1):101–106
48. Lee CY, Lin FL, Yang CT, Wang LH, Wei HL, Tsay HS (1995) Taxol production by cell cultures of *Taxus mairei*. In: Proceedings of symposium on development and utilization of resources of medicinal plants in Taiwan, vol 48. Taiwan Agricultural Research Institute, Taiwan, 21 April, pp 137–148 (TARI Special Publication)
49. Tal B, Rokem JS, Goldberg I (1983) Factors affecting growth and product formation in plant cells grown in continuous culture. Plant Cell Rep 2:219–222
50. Van Hengal AJ, Harkes MP, Witchers HJ, Hesselinic PGM, Buitglaar RM (1992) Characterization of callus formation and camptothecin production by cell lines of *Camptotheca acuminata*. Plant Cell Tiss Org Cult 28:11–18
51. Furuya T, Yoshikawa T, Orihara Y, Oda H (1984) Studies of the culture conditions for *Panax ginseng* cells in jar fermentors. J Nat Prod 47:70–75
52. Hansen G, Wright MS (1999) Recent advances in the transformation of plants. Trends Plant Sci 4:226–231
53. Valluri JV (2009) Bioreactor production of secondary metabolites from cell cultures of periwinkle and sandalwood. Methods Mol Biol 547:325–335
54. Cheetham PSJ (1995) Biotransformations: new routes to food ingredients. Chem Ind 7:265–268
55. Scragg AH (1997) The production of aromas by plant cell cultures. In: Schepier T (ed) Advance biochemical and engineering biotechnology. Springer, Berlin
56. Krings U, Berger RG (1998) Biotechnological production of flavours and fragrances. Appl Microb Biotechnol 49:1–8
57. Verma P, Mathur AK, Srivastava A, Mathur A (2012) Emerging trends in research on spatial and temporal organization of terpenoid indole alkaloid pathway in *Catharanthus roseus*: literature update. Protoplasma 249(2):255–268
58. Lindsey K (1995) Manipulation by nutrient limitation of the biosynthetic activity of immobilized cells of *Capsicum frutescens* Mill. *ev. annum*. Planta 165:126–133
59. Vanisree M, Chen YL, Shu-Fung L, Satish MN, Chien YL, Hsin-Sheng T (2004) Studies on the production of some important secondary metabolites from medicinal plants by plant tissue cultures. Bott Bull Acad Sin 45:1–22
60. Wink M, Alfermann AW, Franke R, Wetterauer B, Distl M, Windhovel J (2008) Sustainable bioproduction of phytochemicals by plant in vitro cultures: anticancer agents. Plant Genetic Resour 12:113–123
61. Jin JH, Shin JH, Kim JH, Chung IS, Lee HJ (1999) Effect of chitosan elicitation and media components on the production of anthraquinone colorants in madder (*Rubia akane* Nakai) cell culture. Biotechnol Bioprocess Eng 4:300–304
62. Zhao J, Hu Q, Guo Q, Zhu WH (2001) Effects of stress factors, bioregulators, and synthetic precursor on indole alkaloid production in compact callus clusters cultures of *Catharanthus roseus*. Appl Microbial Biotechnol 55:693–698
63. Staniszewska I, Krolicka A, Mali E, Ojkowska E, Szafranek J (2003) Elicitation of secondary metabolites in in vitro cultures of *Ammi majus* L. Enzyme Microbiol Technol 33:565–568
64. Jeong GA, Park DH (2006) Enhanced secondary metabolite biosynthesis by elicitation in transformed plant root system: effect of abiotic elicitors. Appl Biochem Biotechnol 129:436–446

65. Minami E, Kuchitsu K, He DY, Kouchi H, Midoh N, Ohtsuki Y, Shibuya N (1996) Two novel genes rapidly and transiently activated in suspension-cultured rice cells by treatment with N-acetylchitoheptaose, a biotic elicitor for phytoalexin production. Plant Cell Physiol 37(4):563–567
66. Jeandet P, Clément C, Courot E, Cordelier S (2013) Modulation of phytoalexin biosynthesis in engineered plants for disease resistance. Int J Mol Sci 14(7):14136–14170
67. Karuppusamy S (2009) A review on trends in production of secondary metabolites from higher plants by in vitro tissue, organ and cell cultures. J Med Plants Res 3:1222–1239
68. Li M, Peebles CA, Shanks JV, San KY (2011) Effect of sodium nitroprusside on growth and terpenoid indole alkaloid production in *Catharanthus roseus* hairy root cultures. Biotechnol Prog 27(3):625–630
69. Shanks JV, Morgan J (1999) Plant hairy root culture. Curr Opin Biotechnol 10:151–155
70. Xu H, Kim YK, Suh SY, Udin MR, Lee SY, Park SU (2008) Deoursin production from hairy root culture of *Angelica gigas*. J Korea Soc Appl Biol Chem 51:349–351
71. Asano T, Kobayashi K, Kashihara E, Sudo H, Sasaki R, Iijima Y, Aoki K, Shibata D, Saito K, Yamazaki M (2013) Suppression of camptothecin biosynthetic genes results in metabolic modification of secondary products in hairy roots of *Ophiorrhiza pumila*. Phytochemistry 91:128–139
72. Flores HE, Hoy MW, Pickard JJ (1987) Secondary metabolites from root culture. Trends Biotechnol 5:64–69
73. Spencer A, Hamill JD, Rhodes MJC (1990) Production of terpenes by differentiated shoot of *Mentha citrate* transformed with *Agrobacterium tummifacience* T-37. Plant Cell Rep 8:601–604
74. Giri A, Narasu ML (2000) Transgenic hairy roots: recent trends and applications. Biotechnol Adv 18:1–22
75. Van Pee KH (2011) Transformation with tryptophan halogenase genes leads to the productionof new chlorinated alkaloid metabolites by a medicinal plant. Chem Biochem 12(5):681–683
76. Kinney AJ (1998) Manipulating flux through plant metabolic pathways. Curr Opin Plant Biol 1:173–178
77. Sato F, Hashimoto T, Hachiya A, Tamura K, Choi KB, Morishige T (2001) Metabolic engineering of plant alkaloid biosynthesis. Proc Natl Acad Sci 2:367–372

# Microbial Pathogenicity: A New Approach to Drug Development

**Ananda M. Chakrabarty**

**Abstract** Pathogenic microorganisms, particularly pathogenic bacteria that cause debilitating diseases, are considered a threat for human health and well-being. Infectious diseases are rampant, particularly in developing countries. However, there are many other deadly diseases, such as cancer, diabetes, cardiac malfunction, etc., that account for significant loss of life and trauma in our society. In this article, I try to describe the role certain bacterial pathogens with long term residence in the human body can play in providing us with our next generation anticancer and other drugs, demonstrating that a certain amount of good can come out of such bacterial pathogens as well.

**Keywords** Microbial pathogenicity · Infectious diseases · Drug development · Cancer

## Introduction

Cancer is a complex disease involving dis-regulation of mammalian cell differentiation and growth. Such cellular events involve cell signaling pathways such as heat shock protein Hsp90, PI3 K/AKT, mTOR, ERK, oncogenes such as Bcl-2 or Ras, and tumor suppressors such as p53 or PTEN, in addition to involvement of growth promoting factors such as vascular endothelial growth factor (VEGF), platelet-derived growth factor (PDGF) and their receptors. The activation of such receptors can in turn activate the Ras/PI3 K/PTEN/AKT/mTOR pathways leading to protein synthesis and other essential cellular processes needed for growth and survival. It is also interesting to note that there is an association between the AIDS

A. M. Chakrabarty (✉)
Department of Microbiology and Immunology, University of Illinois College of Medicine, 835 South Wolcott Avenue, Chicago, IL 60612, USA
e-mail: pseudomo@uic.edu

R. Adhikari and S. Thapa (eds.), *Infectious Diseases and Nanomedicine II*, Advances in Experimental Medicine and Biology 808,
DOI: 10.1007/978-81-322-1774-9_4, © Springer India 2014

virus HIV-1 and cancers including Kaposi's sarcoma, non-Hodgkin's lymphoma, primary brain lymphoma, Burkitt's lymphoma, etc., and inhibitors of mTOR could be useful in the treatment of both AIDS and cancer [1]. The multiple interlinked pathways through which cancer cells grow independent of normal cellular regulation require multiple mutations in the human genome, estimated to be 10 to 100 depending upon the nature and the complexity of the cancer. Such mutations can be acquired during the years, making older people susceptible to cancers. However, such mutations can be induced by external carcinogenic compounds, radiations, certain bacteria and viruses, or can be inherited from parents such as the BRCA1 and BRCA2 mutations that can predispose women to breast and ovarian cancers. Thus pre- or postnatal screenings of such genetic mutations have been an important part of cancer prevention and treatment, although once detected in young adults, such mutations can lead to panic and despair [2].

Once detected in afflicted people, the treatment options for cancer involve surgical removal of the tumor(s) followed by radiation treatment to kill surrounding hidden cancer cells and long term chemotherapy to prevent any residual tumor cells to proliferate. The chemotherapy part is the most devastating part as such agents also kill or affect the growth of normal cells, particularly fast-growing normal cells, producing significant side effects. This is because cancer cells are very much like normal cells, except they escape from normal cellular growth regulation. Aside side-effects, another major problem in anticancer drug usage is the rapid resistance to drugs developed by the cancer cells. Most anticancer drugs are either small molecules or monoclonal antibodies that inhibit a single or a very limited number of targets in the cancer progression pathways. Most cancer cells simply change or stop making this target, using alternate pathways for growth, or use efflux pumps to pump out small molecule drugs, developing drug resistance [3].

## Bacterial Anticancer Drug Development

At present there is no conceivable way current drugs can prevent cancer relapse once the cancer is in remission. Cancer stem cells can hide away from the site of the tumor or new forms of tumor can be triggered by oncogenic transformation of normal cells due to exposure to environmental carcinogens or acquisition of new mutations. Thus to eradicate cancer, a drug, or even a combination of drugs, must not only kill or inhibit the growth of any preformed tumors but must continually prevent the oncogenic transformation of normal cells to cancer cells throughout the lifetime of a person. This requires that the drug or the combination of drugs must not have any significant toxicity, be efficacious and be affordable. Additionally, such drugs should not be amenable to rapid drug resistance. No such drugs or vaccines exist today.

An emerging approach to cancer therapy, besides the current chemotherapy, is to use live attenuated bacteria or viruses and their products [4, 5]. In one particular

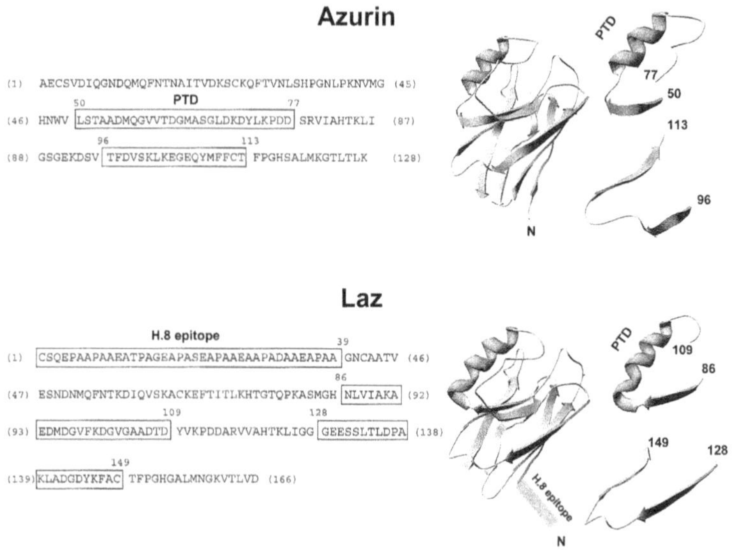


**Fig. 1** Sequence information and structural features of the two proteins Azurin and Laz. Under Azurin, PTD, representing amino acids 50–70, stands for the protein transduction domain. Under Laz, the H.8 epitope sequence is marked [6]

instance, a redox protein termed azurin from the opportunistic pathogen *Pseudomonas aeruginosa*, was shown not only to have anticancer activity but also cancer preventive activity by significantly inhibiting pre-cancerous lesion formation in mouse mammary organ culture (MMOC) where growing normal mammary cells were exposed to a carcinogen 7,12-dimethyl-benz-anthracene, DMBA [6, 7]. Unlike anticancer drugs or monoclonal antibodies that target and inhibit a single or limited number of steps in the cancer growth progression pathway, azurin has the remarkable ability to interfere in multiple pathways by which cancer cells grow involving inhibition of cell signaling pathways and angiogenesis as well as stabilization of tumor suppressors such as p53 [4, 6]. In addition, azurin demonstrated high specificity for entry into cancer cells, but not into normal cells, thus demonstrating an interesting characteristic of distinguishing cancer cells from normal cells and entering preferentially in cancer cells to inhibit by complex formation key intracellular proteins involved in cancer cell growth. The amino acid sequence and structural features of azurin from *P. aeruginosa* and a similar protein Laz, to be described shortly, are shown in Fig. 1 [4, 6].

Will azurin prove effective as an anticancer drug in humans? To address this question, a company, CDG Therapeutics Inc. (CDGTI) in Chicago approached the US FDA to approve azurin's pre-clinical and human clinical trials. Since azurin is a bacterial protein and must be isolated from the bacteria with possible contamination by other toxic bacterial cellular components such as the LPS, the FDA advised the CDGTI to initially try a peptide from azurin that also possesses the

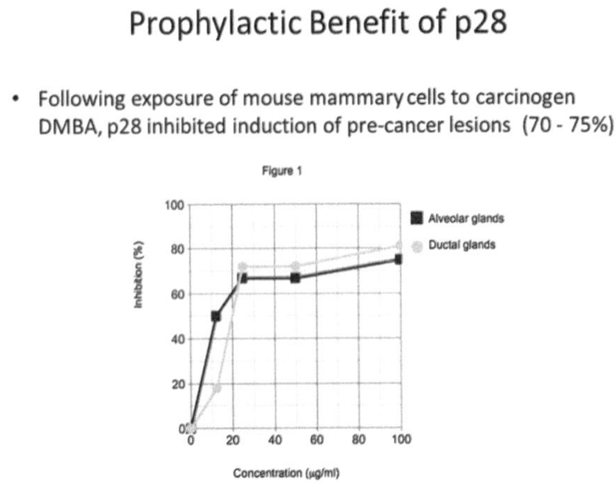


**Fig. 2** Inhibition of Dimethyl-benz-anthracene (DMBA)—induced pre-cancerous lesion formation in mouse mammary organ culture, both alveolar and ductal glands, by increasing concentrations of p28 [7]

anticancer activity and that can be chemically synthesized to be used as a drug. A chemically synthesized 28 amino acid long peptide termed protein transduction domain (PTD; Fig. 1) or p28, with anticancer activity and entry specificity in cancer cells was tested for toxicity in animals including monkeys. No toxicity or immunogenicity was observed. Similar to azurin, p28 had anticancer activity primarily through stabilization of tumor suppressor p53 as well as inhibition of angiogenesis, and also demonstrated cancer preventive properties as evaluated by the MMOC assays in presence of DMBA [2, 7] (Fig. 2). The CDGTI was then approved by the FDA to conduct phase I human clinical trials in 15 advanced stage (stage IV) cancer patients to demonstrate not only its toxicity, or a lack of it, but also any potential beneficial effect. When given as an intravenous bolus in 5 escalating doses in these 15 stage IV patients with refractory metastatic solid tumors (7 melanoma, 4 colon, 2 sarcoma, 1 pancreatic, 1 prostate) three times per week over a 4 week period with a 2 week break before the next, very little immunogenicity or toxicity was observed (Table 1). Of the 15 patients with an average life expectancy of less than 6 months, 2 patients showed partial regression while two patients showed complete regression of their tumors where no conventional drugs were working (Table 1). Of several patients whose life expectancy was prolonged, two patients remain alive disease free at the termination of the trial with their tumors completely regressed [8]. Further, between 2 and 3 years after the trial, three patients remain alive, one with complete regression of the tumors [9].

Why would a peptide derived from a bacterial protein such as azurin demonstrate anticancer activity in drug-resistant cancers without any apparent toxicity or side effects? It should be noted that azurin has not only anticancer activity but it also strongly inhibits host cell invasion by the AIDS virus HIV-1, the malarial parasite *Plasmodium falciparum* and the toxoplasmosis-causing parasite *Toxoplasma gondii* [6]. Thus azurin is believed to be a weapon used by *P. aeruginosa* to keep invaders of the human body in check when these bacteria form biofilms on human epithelial cell surfaces for long term residence without harming or exerting

**Table 1** p28 phase I clinical trials in stage IV cancer patients (metastatic, refractory solid tumors)

| Patients | | | Type of cancer | Treatment modality | Dosage (mg p28/kg) | Adverse events | Objective response |
|---|---|---|---|---|---|---|---|
| Number | 15 | 11 male | Melanoma (7) | i.v. bolus of p28, three times per week for 4 weeks | 0.83<br>1.66 | None attributable to p28 (grade 1) | Stable disease (6/15) Partial tumor regression (2/15, 1 melanoma, 1 prostate) |
| | | 4 male | Colon (4) | | 2.5 | | |
| Age range | 50–80 | | Sarcoma (2) | Observation 2 weeks before the next higher dose | 3.33 | No immune response or adverse effect even on maximum dosage of p28 | Complete regression (2/15, 1 sarcoma, 1 melanoma) |
| Median | 62 | | Pancreatic (1) | | 4.16 in escalating doses | | |
| | | | Prostate (1) | | | | |

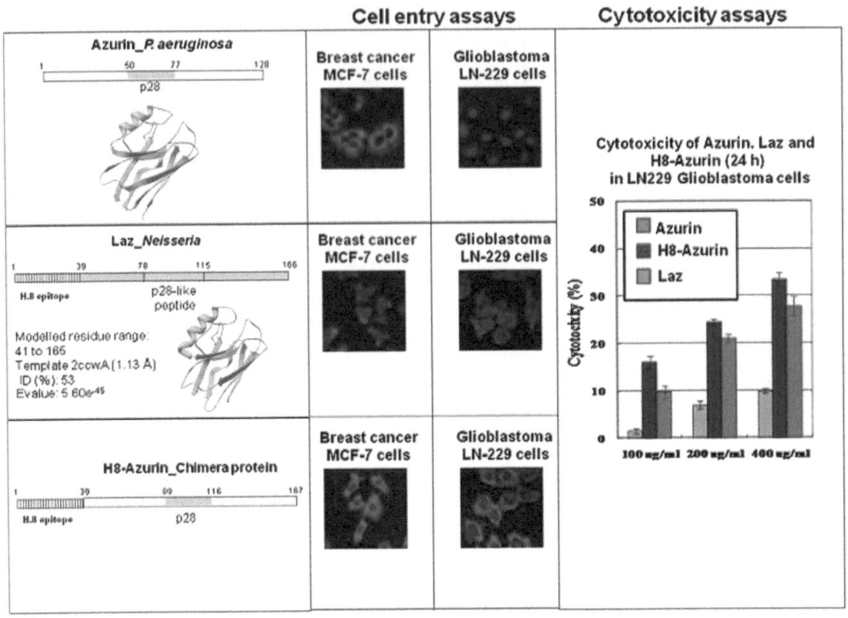


**Fig. 3** *Left row* Structural depiction of azurin, Laz (with the H.8 epitope in the *N*-terminal) and azurin with the cloned H.8 epitope in its N-terminal. The area corresponding to the p28 peptide region is marked. *Middle row* The entry of fluorescently labeled azurin, Laz and H.8 azurin in breast cancer MCF-7 and glioblastoma LN-229 cells is shown. The *red color* reflects the Alexa fluor 568-conjugated protein while the *blue color* represents the nucleus stained with DAPI. *Right row* Cytotoxicity of azurin, H.8-azurin and Laz at three different concentrations towards LN-229 glioblastoma cells [10, 11]

any toxicity to the host [6]. This then raises several questions. Is azurin specific for tumors in the organs where *P. aeruginosa* normally resides during infection? Since *P. aeruginosa* normally does not cross the blood–brain barrier (BBB) to infect the brain meninges, as is the case with *Neisseria meningitides*, the causative agent of meningitis, will azurin be effective against such deadly brain tumors as glioblastomas? Also, if azurin is a weapon produced by a bacterium with long term residence in the human body, are there other protein weapons active against cancers or the HIV/AIDS virus produced by other bacteria with a similar life style?

An insight into the nature of the weapon based on specificity of the bacterial habitat was obtained when we looked at an azurin-like protein produced by *gonococci/meningococci* such as *N. meningitides*. *Neisseria meningitides* produces an azurin-like protein called Laz (lipidated azurin) with a 127 amino acid moiety (*P. aeruginosa* azurin is 128 amino acid long) with 56 % amino acid sequence identity to *P. aeruginosa* azurin. However, Laz harbors an additional 39 amino acid peptide at its N-terminal end called an H.8 epitope (Fig. 3, left column). Interestingly, while azurin is very efficient in entering human breast cancer MCF-7 cells, it was inefficient in entering the brain tumor glioblastoma LN-229 cells. In contrast,

Laz was highly efficient in entering both the breast cancer and the glioblastoma cells (Fig. 3, middle column), suggesting that the presence of the 39 amino acid H.8 epitope in the N-terminal of *Neisserial* azurin aids in its entry to glioblastoma cells. To conclusively demonstrate the role of the H.8 epitope in facilitating the entry of the azurin moiety to glioblastoma cells, we cloned the H.8 epitope in the N-terminal region of *P. aeruginosa* azurin giving rise to the H.8-azurin recombinant protein. In contrast to azurin, H.8-azurin demonstrated efficient entry to the glioblastoma cells (Fig. 3, middle column), suggesting that *N. meningitides* equipped its anticancer weapon azurin with an additional armor, the H.8 epitope, to overcome any entry barrier and to facilitate its weapon's entry to glioblastoma tumor cells to kill them [10].

How does the H.8 epitope work to facilitate such entry to glioblastoma cells and what are the consequences? When the 39 amino acid H.8 epitope was synthesized chemically and incubated with the glioblastoma cells along with azurin, azurin could now enter such cells better than in absence of the H.8 epitope peptide, suggesting that the H.8 epitope destabilizes the glioblastoma endothelial cell tight junctions to facilitate entry of a drug [10], indicating a new way to allow other drugs to enter the brain tumors [11]. This facilitated entry of H.8-azurin and Laz enhances the cytotoxicity of these two proteins towards glioblastoma cells (Fig. 3, right column), demonstrating an interesting role of the H.8 epitope, a unique additional armor added to the weapon azurin by *N. meningitides*, to attack brain tumors when the bacteria are resident in the brain meninges [10, 11].

Glioblastoma multiforme is a highly invasive brain tumor that is normally treated with total surgical resection, followed by combination of radiation therapy and temozolomide [12]. The prognosis for such patients is poor partly because very few drugs can cross the BBB to reach the brain tumor. The high cytotoxicity of H.8-azurin and Laz towards glioblastoma cells, therefore, raised the question if the presence of the H.8 epitope helps azurin, and perhaps the bacteria as Laz is surface-exposed in such bacteria, to cross the BBB. Indeed, when these three proteins were labeled with a green fluorescent dye and injected in the peritoneum of mice, only Laz and H.8-azurin demonstrated significant entry into the brain after 24 h (Fig. 4), showing that the H.8 epitope disrupts the tight junctions in the brain microvasculature endothelial cells to facilitate the entry of *Neisseria*'s anticancer weapon Laz to the brain and attack brain tumors [11, 12].

## Eradicating Cancer in Our Lifetime: Moving Forward

As stated earlier, in order to eradicate cancer in our lifetime, a drug or a combination of drugs, must have at least four important properties. First, the drug should be nontoxic for long term use. The drug also must inhibit the growth of and kill any preformed tumor cells. Thirdly, it must prevent oncogenic transformation of normal cells to cancer cells. These three properties are shared by p28 and probably by azurin, although azurin's lack of toxicity and side effects in humans have not

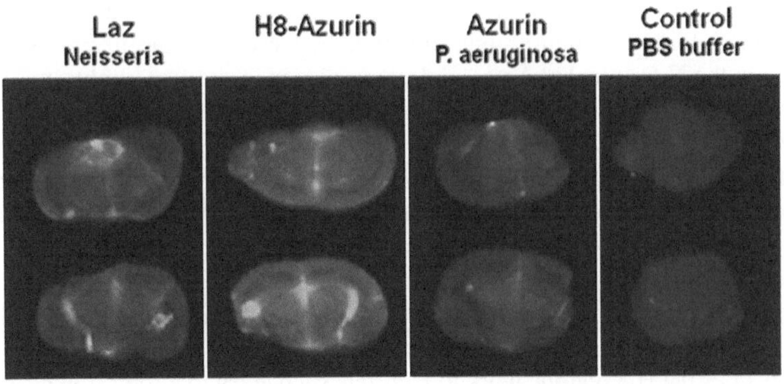


**Fig. 4** Odyssey scanning of brains from mice previously injected peritoneally with *green* fluorescent IR-dye-conjugated Laz, H.8 azurin and azurin (from *P. aeruginosa*). Individual mice were injected with individual fluorescent protein and the mice were sacrificed after 24 h to obtain the brains for Odyssey imaging [11]

yet been assessed. A fourth desirable feature of such a drug is that it should be taken orally and not through intravenous injections. At present, p28 is given intravenously but future technological advancements might make it possible not only to chemically synthesize a protein such as the 128 amino acid long azurin by solid phase peptide synthesis at a modest cost but also to protect such proteins/ peptides from stomach acids and facilitate their gastrointestinal absorption. Current PEGylation (conjugation with polyethylene glycol) techniques can prolong the stability of a protein in serum and reduce immunogenicity and quick renal clearance. If or when such advancements take place, a pill of p28 or azurin a day may keep the cancer away. This could be experimentally demonstrated by giving p28, shown to have very little toxicity in phase I human clinical trials, to vulnerable people predisposed to cancers, such as women harboring BRCA1/BRCA2 mutations, to see if there is statistical reduction in the appearance of breast/ovarian tumors in women taking p28, as compared to a control group that does not [2]. Another important question is if azurin is the only protein produced by *P. aeruginosa* as an anticancer weapon or if there are other proteins, produced by other bacteria with the ability to cause chronic infections and have long term residence in human bodies, to defend the body from invaders such as cancers, viruses and parasites. It is thus interesting to note that azurin is not the only anticancer agent produced by *P. aeruginosa*. Arginine deiminase (ADI), a protein known to have anticancer and anti-viral activity, from *P. aeruginosa* has also been shown to have strong anticancer activity against leukemia, breast and ovarian cancers [5]. In addition, an Indian company Amrita Therapeutics (http://www.amritatherapeutics.com) has recently reported the isolation of a protein ATP-01 from *Mycobacteria* with both anticancer and anti-HIV/AIDS activity. Similar to p28, a 30 amino acid chemically synthesized peptide AT-01, derived from ATP-01, shows anticancer activity against skin, colon and other cancers [12]. It remains to be seen if other pathogenic or even nonpathogenic bacteria may produce other proteins, somewhat similar to antibiotics, with anticancer, anti-viral and anti-parasitic activities, giving rise to a new type of industry rivaling the antibiotic industry. If the pathogenic bacteria with long term residence in human body produce these proteins to defend their habitat from invaders such as cancers, viruses and parasites, it is also conceivable that such bacteria may produce similar protein weapons that can protect the human body from other deadly diseases or their agents such as coronary thrombosis or diabetes. It will thus be exciting to look for bacterial proteins that can not only potentially eradicate cancer but that can also form complexes with nuclear receptors, for example thyroid receptors or peroxisome proliferator-activated receptor gamma that are known to be involved in obesity, diabetes or atherosclerosis, to keep such diseases in check, heralding a new era of drug development and contributing to better human health.

## References


1. Nicoletti F, Fagone P, Meroni P, McCubrey J, Bendtzen K (2011) mTOR as a multifunctional therapeutic target in HIV infection. Drug Discov Today 16:715–721
2. Fialho AM, Chakrabarty AM (2012) Patent controversies and court cases: Cancer diagnosis, therapy and prevention. Cancer Biol Ther 13:1229–1234
3. Avner BS, Fialho AM, Chakrabarty AM (2012) Overcoming drug resistance in multi-drug resistant cancers and microorganisms: a conceptual framework. Bioeng Bugs 3:262–270
4. Bernardes N, Chakrabarty AM, Fialho AM (2013) Engineering of bacterial strains and their products for cancer therapy. Appl Microbiol Biotechnol. doi:10.1007/s00253-013-4926-6
5. Fialho AM, Chakrabarty AM (eds) (2010) Emerging cancer therapy: microbial approaches and biotechnological tools. Wiley, Hoboken
6. Fialho AM, Chakrabarty AM (2010) Promiscuous anticancer drugs from pathogenic bacteria: rational versus intelligent drug design. In: Fialho AM, Chakrabarty AM (eds) Emerging cancer therapy: microbial approaches and biotechnological tools. Wiley, Hoboken, pp 181–198
7. Das Gupta TK, Chakrabarty AM (2009) Compositions and methods to prevent cancer with cupredoxins. US patent 7,618,939, 17 Nov 2009
8. Richards JM, Warso MA, Mehta D, Christov K, Schaeffer CM, Yamada T, Beattie CW, Bressler LR, Das Gupta TK (2011) A first-in-class, first-in-human phase I trial of p28, a non-HDM2-mediated peptide inhibitor of p53 ubiquitination in patients with metastatic refractory solid tumors. J Clin Oncol 29:2511 (abstract)
9. Warso MA, Richards JM, Mehta D, Christov K, Schaeffer C et al (2013) A first-in-class, first-in-human, phase I trial of p28, a non-HDM2-mediated peptide inhibitor of p53 ubiquitination in patients with advanced solid tumours. Br J Cancer. doi:10.1038/bjc.2013.74
10. Hong CS, Yamada T, Hashimoto W, Fialho AM, Das Gupta, TK, Chakrabarty AM (2006) Disrupting the entry barrier and attacking brain tumors: the role of the Neisseria H.8 epitope and the Laz protein. Cell Cycle 5:1633–1641
11. Hong CS, Yamada T, Fialho AM, Das Gupta, TK, Chakrabarty AM (2010) Transport agents for crossing the blood-brain barrier and into brain cancer cells, and methods of use thereof. US patent 7,807,183, 5 Oct 2010
12. Fialho AM, Salunkhe P, Manna S, Mahali S, Chakrabarty AM (2012) Glioblastoma multiforme: novel therapeutic approaches. ISRN Neurology: 642345; PMID: 22462021. doi: 10.5402/2012/642345

# Current Trend on Syphilis Diagnosis: Issues and Challenges

**Muhammad G. Morshed**

**Abstract** Syphilis is a century old sexually transmitted infection transmitted worldwide. WHO reports 12 million new cases that were identified in 1999, with over 90 % of these infections reported from patients from low-income countries. This case number of syphilis is on the rise globally in the Men have sex with Men (MSM) population. Dark field microscopy (DFM) and direct fluorescence assay (DFA) have been used in clinical laboratories for decades to demonstrate *Treponema pallidum* in acutely infected human tissue and/or body fluids. Molecular technologies allow detecting *T. pallidum* and also determine drug resistance (by identifying DNA point mutation). It is evident from the published literature that PCR is useful as an adjunct test to DFA and DFM, and is useful in confirming syphilis in genital ulcer, tissue, and other body fluid samples, providing even more sensitive detection algorithm. Serological tests remain the mainstay tests since *T. pallidum* is nonculturable and also because blood collection is easy. The practice of serological testing is changing rapidly from traditional nontreponemal screening followed by confirmatory treponemal testing to screening by treponemal tests referred to as "Reverse Algorithm" followed by nontreponemal testing to determine active infections. Special and further complex algorithms are essential to deal with complex issues such as neurosyphilis or congenital syphilis. Due to the huge surge of syphilis in developing countries where access to medical care is not optimal, point of care or rapid tests may play an important role. This author took an attempt to summarize the current trend of syphilis diagnosis and challenges from a global perspective.

**Keywords** Syphilis · Treponema pallidum · Serology · Diagnosis · Detection

M. G. Morshed (✉)
Department of Pathology and Laboratory Medicine, Faculty of Medicine, University of British Columbia; BC Public Health Microbiology and Reference Laboratory, Provincial Health Services Authority, 655 West 12th Avenue, Vancouver, BC V5Z 4R4, Canada
e-mail: muhammad.morshed@bccdc.ca

R. Adhikari and S. Thapa (eds.), *Infectious Diseases and Nanomedicine II*, Advances in Experimental Medicine and Biology 808,
DOI: 10.1007/978-81-322-1774-9_5, 

## Introduction

Syphilis is a centuries old disease caused by the bacterium *Treponema pallidum* sub spp *pallidum*. It was first reported in Europe (Italy) at the end of the fifteenth century [1]. Infections are mostly sexually transmitted as well as spread from an infected mother to her fetus or through blood transfusions [2] and rarely through organ transplant [3].

While *T. pallidum* remains highly sensitive to penicillin, it remains a worldwide scourge. Globally, 25 million people are infected, with an estimated annual incidence of 12 million cases [4]. Syphilis rates have been increasing in several global jurisdictions over the past two decades. In British Columbia and Cannada the number of syphilis cases have increased alarmingly since 2000 and in 2102 the rate was the highest recorded in the last 30 years (Fig. 1) [5, 6].

*T. pallidum* can cause disease in three stages. Entering through intact or abraded skin or mucous membrane and multiplying at the site of entry, this spirochete may cause a usually painless ulcer or ulcers in approximately 3 weeks' (range 9–90 days) postexposure; this is the first stage of syphilis. Without treatment infections may resolve within 1–5 weeks [7, 8].

Humoral antibodies against cardiolipin (a nonspecific antigen that was discovered to cross-react with *T. pallidum* antigens) and treponemal antigen (from animal sources) usually do not appear until 1–4 weeks after the chancre. During the time the ulcer heals, *T. pallidum*, when not stopped by treatment, spreads systemically and may cause signs and symptoms; this is the second stage of syphilis.

Multiple types of rashes and flu-like symptoms mimicking many other diseases may appear about 2–6 weeks later. This secondary syphilis stage, if untreated, resolves within 2–6 weeks. The infection can go on to the third stage (the tertiary stage), 3–30 years later. Symptoms are chronic manifestations of disease including gummata (chronic granulomas in any tissue) and cardiovascular or neurological signs and symptoms [8].

A laboratory diagnosis is an important tool to confirm infection for patients who have been exposed to *T. pallidum*. For positive patients, serial Rapid Plasma Regain (RPR) tests help to monitor treatment progress as well as able to determine reinfection.

## Laboratory Diagnosis of *Treponema pallidum*

### *Direct Detection of* Treponema pallidum

In cases where syphilitic ulcer and/or rashes are evident, direct detection of treponema tests may be performed to diagnose the infection when the lesions are suitable.

Methods for the direct detection of syphilis include rabbit infectivity testing (RIT), dark field (DF) or Direct Fluorescent Antibody (DFA), Histology followed

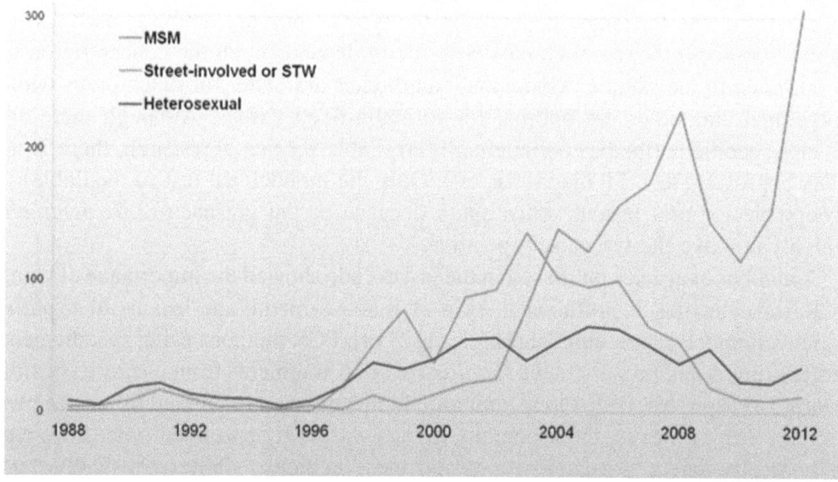


**Fig. 1** Infectious syphilis cases by exposure category in British Columbia, 1988–2012 (*Source* BC Centre for Disease Control. *Courtesy* Dr. Richard Lester and Travis Hottes)

by silver staining, and most recently, the polymerase chain reaction (PCR). The PCR is now becoming a popular tool and has advantages over other direct detection testing.

Rabbit infectivity testing (RIT) is one of the oldest methods and has been considered as the gold standard test for the direct detection of *T. pallidum*. Briefly, samples need to be injected in rabbit testes within an hour to have the optimal outcome. This test can be used for a wide variety of clinical samples (blood, cerebrospinal fluid (CSF,) amniotic fluid, primary and secondary lesion exudates, and lymph node (LN) aspirate). The technique is highly sensitive and very specific and is said to be capable of detecting as few as 1–2 organisms. However, it is impractical to use this technique in a clinical laboratory [9].

Dark Field (DF) microscopy is another old technology but still works well if laboratory and clinics are situated in close proximity. This technology visualizes *T. pallidum* based upon its unique characteristic morphology and motility. It can be used for moist primary and secondary lesions, exudates, LN aspirate, CSF, amniotic fluid, and other body fluids. The sensitivity of DF microscopy ranges from 74 to 79 %, however, it depends on the state of lesion and at least $10^4$–$10^5$ spirochetes need to be present per milliliter of sample. Specificity is highly dependent on the skill of the laboratory technologists, state of fluid collected, transition time from clinic to laboratory and ongoing training of lab staff and monitoring with quality assurance programs [10]. It is not of use for oral lesions because of the similar morphology of *T. denticola*.

Immunostaining identifies *T. pallidum* based on antigen detection and morphology. DFA-TP (direct fluorescent antibody staining for *T. pallidum*), is an immunofluorescence enzyme-based microscopy method that can be used for lesion smears, concentrated fluids, tissue brushings, and fixed or unfixed tissues [11]. The specificity of the technique depends on the type of primary antibody used (polyclonal, monospecific, monoclonal). Specificity depends upon the concentration of *T. pallidum* in the sample. Monoclonal antibodies are better for touch preps while polyclonal antibodies are optimal for formalin-fixed tissues. Although there are several specific antibodies commercially available for use in research, there is no FDA approved DFA-TP diagnostic test. Only the monoclonal test (if available) is acceptable for oral lesions, once again because of the presence of *T. denticola* which can make the test falsely positive.

A number of articles published in the last decade showed the importance of using PCR for detecting *T. pallidum* in skin or mucous membrane lesions of syphilis patients found as a best clinical samples [12–14]. PCR methods using specific gene target primers and probes detect *T. pallidum* DNA fragments from ulcerative lesion swabs, LN aspirates, CSF, blood, amniotic fluid, and fixed or unfixed tissue samples [12, 15, 16]. However, the specificity greatly varied by specimen type. PCR can theoretically detect 1–2 gene copies and the specificity, while intrinsically very high, depends on primer selection, skill of the laboratory technicians, sample type, quality, and handling [17]. A few commercial tests are available globally but need to be validated properly before implementation. In-house testing works well and results need to be aligned with serological findings as well as clinical presentations. PCR is not routinely used for diagnosis of syphilis; however, experience has shown that PCR-based methodologies have distinct advantages, as an adjunct test, over DF and DFA-TP. First, PCR is an effective test for oral lesions obviating the need for the DFA-TP test, which uses a monoclonal that is no longer available, or invasive biopsy procedures. Second, the sequence analysis of the DNA of *T. pallidum* assists in molecular epidemiology of syphilis. Finally, the PCR identifies macrolide resistance and facilitates optimal treatment [18, 19].

## Serological Diagnoses of *T. pallidum* Infection

Since *T. pallidum* cannot be cultured, antibody produced in response to syphilis infection is another way to confirm infection. Because of the ease of blood collection compared to taking samples from lesions, serology becomes the mainstay test for diagnosing syphilis infection. Serology tests can be divided into nontreponemal test and treponemal tests. Nontreponemal tests measure Immunoglobulins G (IgG) and IgM antibodies formed by the host in response to lipoidal material released from damaged host cells. It also considered that some cardiolipin is released by the spirochetes as well. There is no fixed protocol and each laboratory must determine the algorithm based on their population, lab infrastructure, needs, and financial considerations [20].

Treponemal tests use whole *T. pallidum* or its components as antigen sources. A few purified proteins have been used on recently developed Enzyme Immuno Assays (EIA) and Point of Care (POC) tests which will be discussed elsewhere in

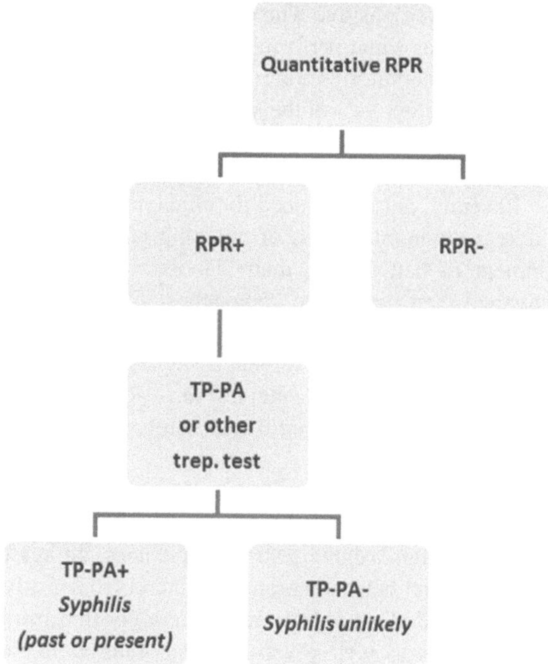


**Fig. 2** Traditional syphilis serology algorithm recommended by CDC, USA. *RPR* Rapid Plasma Reagin, *TPPA Treponema pallidum* Particle Agglutination Test

this article. The use of only nontreponemal or only treponemal tests in isolation is not ideal since each test has limitations [21].

Traditionally, syphilis serology screening tests started with a nontreponemal test such as the rapid plasma regain (RPR) or VDRL; when positive, results were confirmed using specific treponemal tests such as *Treponema pallidum* Particle Agglutination (TPPA) or fluorescent antibody-absorption (FTA-Abs) Fig. 2.

Certain laboratories may wish to use an RPR cutoff equal or above 1:8 and may choose not to do TPPA test if supported by their data (Dr. Bushra Shaihr, Personal Communication); however, these options are the choice of each laboratory which may need to cater test batteries according to their needs and resources. Since RPR positive titers can correlate with disease activity, they can be useful for monitoring treatment or reinfection. The RPR test is comparatively cheap and can be performed in nearly any laboratory setting. A refrigerator is required to keep the reagents cool. However, traditional RPR-based approaches have a few major issues. Some old treated or untreated cases will be missed if using this algorithm because in some instances the RPR will be negative. The RPR test may be reactive from areas where yaws, pinta, or non-venereal syphilis disease is endemic. False positive reactions may also occur mainly in samples of patients who have other diseases such as lupus, mononucleosis, malaria, leprosy, viral pneumonia, rickettsial infection, etc. False positive patterns, e.g., RPR Negative, TPPA reactive can occur with very early syphilis infection and few latent syphilis infections [22]. On the other hand, the prozone reaction could give false negative RPR results which become positive when the serum is diluted out and technologists should be aware of that when performing RPR tests [23].

Certain laboratories prefer to use the VDRL test as screening test. However, most of the laboratory uses of the VDRL are for testing CSF samples in order to diagnose neurosyphilis. Sensitivity of VDRL test for neurosyphilis is not optimal. Thus, further tests need to be performed if neurosyphilis is clinically suspected [24, 25].

Recently, due to the need for efficiencies in high-volume screening, the need to address ergonomic stress of pipetting large numbers of samples, and with development of better tests, many laboratories in Europe and North America have changed their diagnostic approaches. They screen using automated or semiautomated Treponemal antibody assays in which the blood sample is tested using an enzyme-linked immunosorbent assay (EIA). Positive samples are then tested with a quantitative nontreponemal test (e.g., RPR or VDRL). If any test results disagree, the specimen is also then tested using the TPPA test as the confirmatory treponemal test (Fig. 3).

If TPPA is 1+ or equivocal, some laboratory may use another test as a tie breaker such as INNO-LIA to confirm the diagnosis. This new screening approach results in high-volume testing efficiencies, addresses ergonomic issues for technologists, and detects more cases (increased sensitivity) for early as well as latent syphilis [26]. This reverse sequence algorithm has been adopted by many of the large private as well government laboratories in Europe and North America and almost all of them are using fully automated systems. Available systems are offered by Siemens, Abbott, BioRad, to name a few.

One challenge with this new approach, however, is that there are more false positives due to increased test sensitivity [27, 28]. Upfront setup cost as well as recurrent laboratory cost may be increased as well [29, 30]. Ideally, a screening test should be simple, easy to use, provide rapid results to enhance faster therapeutic interventions, and have the sensitivity, specificity, positive, and negative predictive values suitable for use in both low- and high-prevalence populations. It also needs to be cost-effective. Meeting all these testing criteria is not simple.

Irrespective of either algorithm, it is important that the clinician should always consider all other factors such as sexual behavior, medical history, and previous treatment history of syphilis, etc. If the clinical assessment suggests possible or probable syphilis and the syphilis screening test is nonreactive (traditional algorithm), the laboratory test should be repeated after 2–4 weeks. If the RPR is still nonreactive, early syphilis is unlikely and latent syphilis is the most likely diagnosis. These patients should be treated in case they could go on to develop tertiary syphilis.

## Complex Syphilis Case Diagnosis

Neurosyphilis and congenital syphilis need different algorithms to diagnose these extremely complex cases.

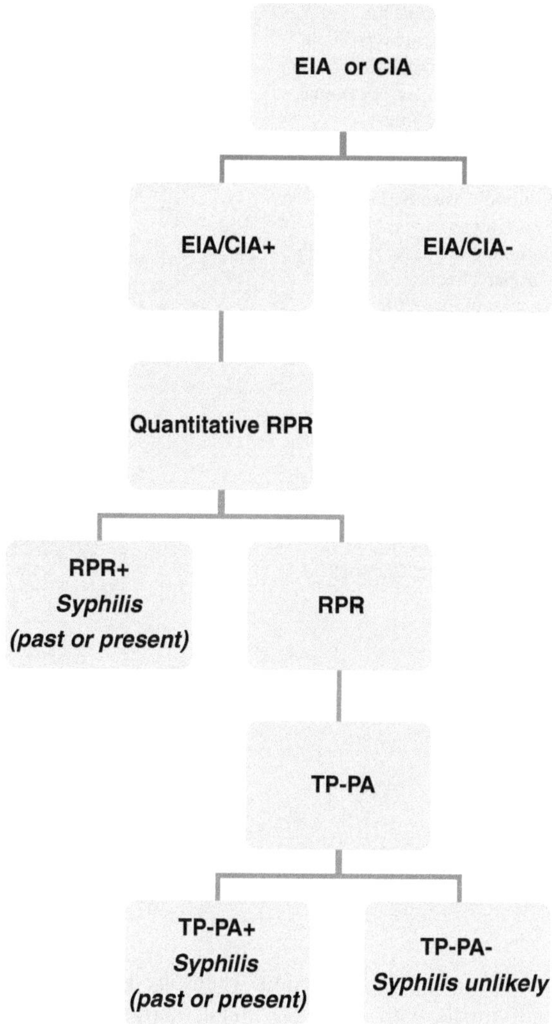


**Fig. 3** Reverse sequence serology algorithm recommended by CDC, USA. *EIA* Enzyme Immunoassay, *CIA* Chemiluminescence Immunoassay, *RPR* Rapid Plasma Reagin, *TPPA* *Treponema pallidum* Particle Agglutination Test

## *Neurosyphilis*

Neurosyphilis was a more common phenomenon after *T. pallidum* infection in the preantibiotic era [31, 32]. However, currently the percentage is still high in certain group of patients, e.g., Men have sex with Men (MSM) patients with HIV infection [33]. It happens in people who are not HIV positive as well, so the clinician must be on the alert for this.

To make it easier for laboratory scientists and clinicians, the neurosyphilis algorithm can be divided into two; one is without HIV infection (Fig. 4) and the

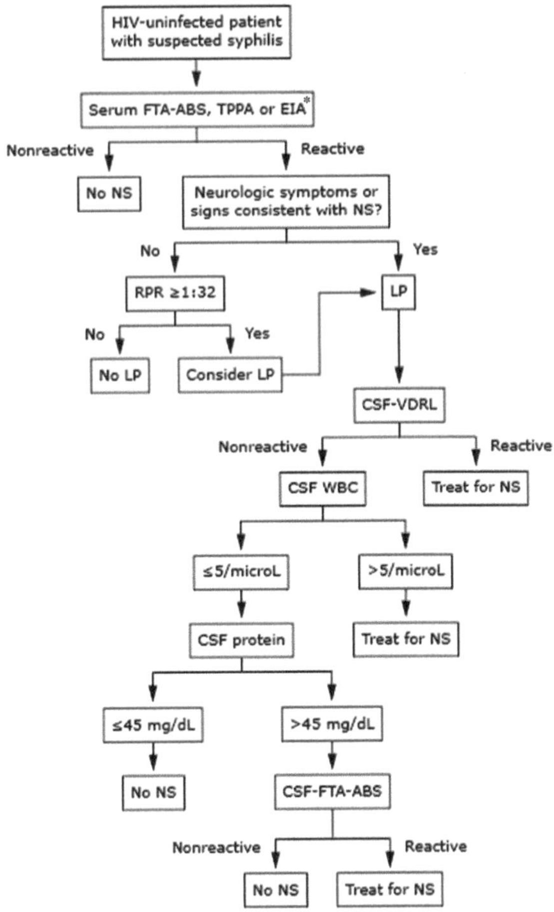


**Fig. 4** Algorithm for diagnosis of neurosyphilis in a patient without HIV infection. *CSF* Cerebrospinal Fluid, *EIA* Enzyme Immunoassay, *FTA-Abs* Fluorescent Treponemal Antibody-Absorbed Test, *HIV* Human Immunodeficiency virus; *LP*: Lumbar Puncture; *NS*: Neurosyphilis; *RPR*: Rapid Plasma Reagin test; *TPPA*: *Treponema pallidum* Particle Agglutination test; *VDRL*: Venereal Disease Research Laboratory test; *WBC*: White Blood Cell Count. *A reactive serum EIA should be confirmed with a reactive serum FTA-Abs or TPPA. *Courtesy of* Dr. Christina M Marra, MD

other is with HIV infection. There are two tables because (Fig. 5) HIV infected individuals with low CD4 counts showed different serologic and CSF findings compared to an immunocompetent host representing neurosyphilis without HIV infection in this article.

Often the microbiology laboratory does not get any clinical information on the requisition form except the test request as neurosyphilis, neuritis, or uveitis. Neither do they get other test results such as CD4 counts or protein concentration in CSF and it is not usually practical to call each ordering physician to find out the clinical information, as well as other clinical chemistry and cell count results. In such situations, it is better for any microbiology laboratory to do all available treponemal serology such as EIA, TPPA, and FTA-Abs using CSF samples even if

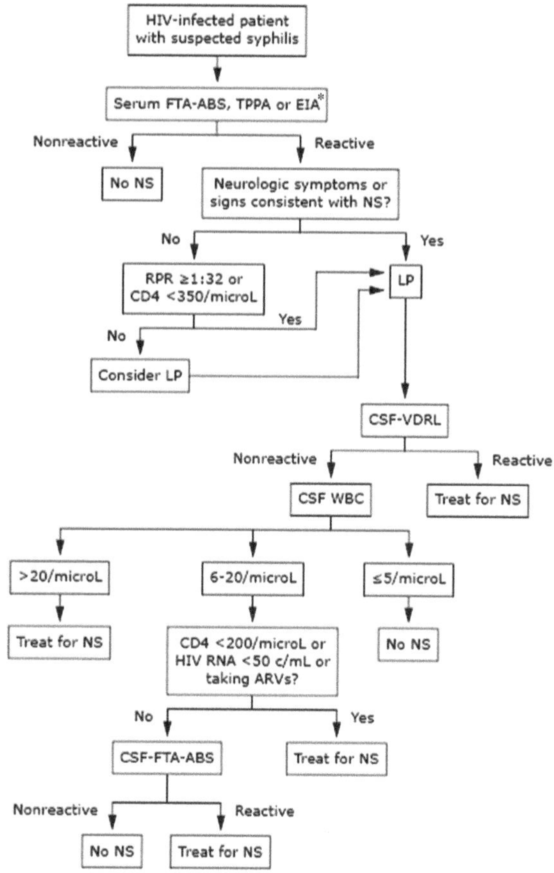


**Fig. 5** Algorithm for diagnosis of neurosyphilis in a patient with HIV infection. *ARV* Antiretroviral Drugs, *CD4* Cluster Determinant 4 T Lymphocyte, *CSF* Cerebrospinal Fluid, *EIA* Enzyme Immunoassay, *FTA-Abs* Fluorescent Treponemal Antibody-Absorbed Test, *HIV* Human Immunodeficiency virus, *LP* Lumbar Puncture, *NS* Neurosyphilis, *RNA* Ribonucleic acid, *RPR* Rapid Plasma Reagin test, *TPPA Treponema pallidum* Particle Agglutination Test, *VDRL* Venereal Disease Research Laboratory Test, *WBC* White Blood Cell Count. *A reactive serum EIA should be confirmed with a reactive serum FTA-Abs or TPPA. *Courtesy of* Dr. Christina M Marra, MD

VDRL is negative. If any of these have a positive signal then the ordering physician should do a careful clinical assessment keeping serology results and clinical presentation in mind.

## Test for Congenital Syphilis

Congenital syphilis is infection of the fetus or newborn child. Usually it occurs when *T. pallidum* is transmitted from the infected mother to her fetus. This may result in stillbirth, premature delivery, and a wide range of clinical manifestations to the newborn infant. There also can be later manifestations (after 2 years… Late

Congenital Syphilis). Congenital syphilis is a significant global public health problem and approximately 1 million mothers are affected throughout the world per year [34, 35]. If any newborn child is being considered as a suspected congenital syphilis case (infant born to mother with infectious syphilis without treatment or partial treatment or treatment within 4 weeks of delivery) both mother and infant need to be worked up extensively. In the case of the mother, both treponemal and nontreponemal tests should be performed on her serum. If during examination of the mother a rash or ulcerative lesion(s) are found, direct tests such as PCR or DFA should be performed if the lesions are moist. If the mother is found to be positive for syphilis, an extensive work-up is needed for the infant including serology, radiology, blood CBC, CSF cell count, and protein as well as CSF VDRL [36]. If there is a nasal discharge, PCR or DFA should be performed on the fluid.

## Point of Care Test (POC)

The World Health Organization (WHO) has promoted the syphilis Point of Care Test (POCT) in order to reduce the syphilis case burden in developing countries and specifically for African nations. POCT is also called "rapid test for syphilis" and it may be defined is a test that can be performed outside of a laboratory setting for which the result is available without reference to a laboratory and is rapid enough to result in immediate patient management [37]. For developing countries, the tests should be affordable, sensitive, specific, user friendly, and simple to perform with minimal training, rapid, robust, equipment free, and delivered to those who need it. Many developed countries have licensed POCT for uses in specific populaces. For example, in Canada the test has the potential for use in hard to reach, transient high risk populations where traditional venipuncture approaches are impractical. At the beginning, the POCT that was available could test only for treponemal antibodies. Now companies have developed a test that can determine both treponemal and nontreponemal antibodies. This POCT can definitely help to determine active versus past infection. Sensitivity and specificity for these tests need further evaluation. Several studies have been carried out to determine the sensitivity and specificity of treponemal antibody detection tests and result were varied based on the sample type, study population, staff skill level with the test and interpretation, etc. (Table 1).

**Table 1** Published studies of syphilis POCT in antenatal and STI clinics (adapter from Ref. [38])

| | Location; study population | Test name | Reference standard test | Samples | Sensitivity (95 % CI) | Specificity (95 % CI) | Syphilis antibody prevalence |
|---|---|---|---|---|---|---|---|
| Bronzan, 2007 (finger prick) | South Africa, eight rural clinics | Determine | RPR, TPHA | 341 | 86 (57–98) | 91 (87–94) | 6.5 |
| Hernandaz-Trejo, 2006 (serum) | Mexico, two urban clinics | Determine | VDRL, FTA-Abs | 1322 | 100 | 100 | 0.3 |
| Nessa et al. | Dhaka Bangladesh | Syphilis ultra rapid (ACON, USA) | RPR, TPHA | 684 | 94.4 | 92.6 | 20–76 |
| Mabey, 2006 (multiple) | Tanzania, one large government clinic | Determine | two rapid tests compared with treponemal tests (TPPA, etc.) | 528 | 60 (47–72) | 99 (99–100) | 10.8 |
| | | Visitect | | 528 | 75 (64–86) | 99 (99–100) | 10.8 |
| | | Syphicheck-WB | | 582 | 79 (68–89) | 99 (98–100) | 9.45 |
| | | SD Bioline 3.0 | | 582 | 86 (77–94) | 98 (97–99) | 11.34 |
| Montoya, 2006 (finger prick) | Mozambique, six rural clinics | SD Bioline 3.0 | RPR, TPHA | 326 | 86 (82-89) | 97 (96-97) | 8.35 |
| Tinajeros, 2006 (finger prick) | Bolivia, four large urban clinics | Determine | RPR, TPPA | 8892 | 92 (88-94) | 98 (98-99) | 3.85 |
| Villazon-Vargas, 2009 (whole blood; not finger prick) | Bolivia, one urban clinic | Determine | RPR, FTA-Abs | 489 | 98 | 100 | 4.5 |
| West, 2002 (serum) | Gambia, one rural clinic | Rapid syphilis test | RPR, TPHA | 1325 | 75 | 95 | 3.0 |

## Conclusions

Sample selection and timely testing is key for any diagnostic test and the same principle applies to syphilis. An algorithmic approach usually works better than a random test. Occasionally, serial tests are needed in order to confirm the diagnosis as to whether it is early infectious syphilis or syphilis of longer duration. Traditional RPR-based serology works well if done in the proper settings. Reverse algorithm screening with treponemal tests can be easily automated and has potential to be diagnosed very early as well as latent infection. Molecular tests should be considered for direct detection in conjunction with DF and DFA, or in the case of oral lesions where these are the tests of choice. POC test should be done in a population where traditional testing is not practical.

**Acknowledgments** Thanks are due to Dr. Hugh Jones MD, Diploma in Venerology (UK) for his review and critical reading of this manuscript; Dr. Christina Marra for providing neurosyphilis and congenital syphilis algorithms and Dr. Richard Lester for providing Canadian syphilis data.

## References

1. Farhi D, Dupin N (2010) Origins of syphilis and management in the immunocompetent patient: facts and controversies. Clin Dermatol 28:533–538
2. Rockwell DH, AE Yobs, Moore JB Jr (1964) The Tuskegee study of untreated syphilis; the 30th year of observation. Arc Intern Med 114:792
3. Cortes NJ, Afzali B, MacLean D et al (2006) Transmission of syphilis by solid organ transplantation. Am J Transplant 6:2497–2499
4. Gerbase AC, Rowley JT, Heymann DH et al (1998) Global prevalence and incidence estimates of selected curable STDs. Sex Transm Infect 74(Suppl1):S12–S16
5. Ogilvie GS, Taylor DL, Moniruzzaman A et al (2009) A population-based study of infectious syphilis rediagnosis in British Columbia, 1995–2005. Clin Infect Dis 48:1554–1558
6. Lester RT, Morshed M, Gilbert M, Hottes T (2013) Syphilis and Neurosyphilis increase to historic level in BC. BCMJ 55(4):204
7. Morshed M (2012) Better syphilis infection detection in better patient care and disease prevention. BCMJ 54(6):306–307
8. Larsen SA, Johnson RE (1998) Diagnostic test. In: Larsen SA, Pope V, Johnson RE, Kennedy EJ Jr (eds) A manual of tests for syphilis (1998), 9th edn. American Public Health Association 9:1–52
9. Magnuson H, Eagle H, Fleischman R (1948) The minimal infectious inoculum of *Spirochaeta pallida* (Nichols strain) and a consideration of its rate of multiplication in vivo. Am J Syph 32:1–18
10. Kennedy EJ, Creighton ET (1998) Dark field microscopy for the detection and identification of *Treponema pallidum*. In: Larsen SA, Pope V, Johnson RE, Kennedy EJ Jr (eds) A manual of tests for syphilis, 9th edn. American Public Health Association 9:119–133
11. George RW, Hunter EF, Fears MB (1998) Direct fluorescent antibody test for *Treponema pallidum* (DFA-TP) In: Larsen SA, Pope V, Johnson RE, Kennedy EJ Jr (eds) A manual of tests for syphilis, 9th edn. American Public Health Association 9: 135–146
12. Grange PA, Gressier L, Dion PL et al (2012) Evaluation of a PCR test for detection of *Treponema pallidum* in swabs and blood. J Clin Microbiol 50:546–552

13. Heymans R, van der Helm JJ, de Vries HJC, Fennema HSA, Coutinbo RA, Bruisten SM (2010) Clinical value of *Treponema pallidum* real-time PCR for diagnosis of syphilis. J Clin Microbiol 48:497–502
14. Palmer HM, Higgins SP, Herring AJ, Kingston MA (2003) Use of PCR in the diagnosis of early syphilis in the United Kindgom. Sex Transm Infect 79:479–483
15. Morshed MG, Lee MK, Mguire J et al (2008) Neurosyphilitic gumma in a homosexual man with HIV infection confirmed by polymerase chain reaction. Int J STD AIDS 19(8):568–569
16. Wicher K, Noordhoek GT, Abbruscato F, Wicher V (1992) Detection of *Treponema pallidum* in early syphilis by DNA amplification. J Clin Microbiol 30:497–500
17. Martin IE, Tsang RSW, Sutherland K et al (2009) Molecular characterization of syphilis in patients in Canada: azithromycin resistance and detection of *Treponema pallidum* DNA in whole-blood samples versus ulcerative swabs. J Clin Microbiol 47:1668–1673
18. Lukehart SA, Godomes C, Molini BJ et al (2004) Macrolide resistance in *Treponema pallidum* in the United States and Ireland. N Engl J Med 351:154–158
19. Matejkova P, Flasarova M, Zakoucha H et al (2009) Macrolide treatment failure in a case of secondary syphilis: a novel A2059G mutation in the 23S rRNA gene of *Treponema pallidum* subsp. *pallidum*. J Med Microbiol 58:832–836
20. Tsang ESW, Radons SM, Morshed M (2011) Laboratory Diagnosis of syphilis: a survey to examine the range of tests used in Canada. Can J Infect Dis Med Microbiol 22(3):83–87
21. Workowski KA, Berman S (2010) Sexually transmitted diseases treatment guidelines. Centre for Disease Control and Prevention (CDC). Recommendation and Reports 59 (RR12):1–110
22. Larsen SA, Steiner BM, Rudolph AH (1995) Laboratory diagnosis and interpretation of tests for syphilis. Clin Microbial Rev 8:1–21
23. Berkowitz K, Baxi L, Fox HE (1990) False-negative syphilis screening: the prozone phenomenon, non immune hydrops, and diagnosis of syphilis during pregnancy. Am J Obstet Gynecol 163(3):975–977
24. Larsen SA, Hambie EA, Wobig GH, Kennedy EJ (1986) Cerebrospinal fluid serologic tests for syphilis: treponemal and non-treponemal tests. In: Morriset R, Karstak E (eds). Advances in sexually transmitted diseases. VNW Science Press, Utrecht, pp 157–162
25. Kennedy EJ, Creighton ET (1998) Veneral Disease Research Laboratory (VDRL) slide tests In: Larsen SA, Pope V, Johnson RE, Kennedy EJ Jr (eds) A manual of tests for syphilis, 9th edn. American Public Health Association 9:159–177
26. Misra S, Boily M-C, Ng V et al (2011) The laboratory impact of changing syphilis screening from the rapid-plasma regain to a treponemal enzyme immuno assay: a case study from the greater Toronto area. Sex Transm Dis 38:190–196
27. CDC (2008) Syphilis testing algorithms using treponemal tests for initial screening—four laboratories, New York City, 2005–2006. MMWR 57:872–875
28. CDC (2011) Discordant results from reverse sequence syphilis screening—five laboratories, United States, 2006–2010. MMWR 60:133–136
29. Chuck A, Ohinmaa A, Tilley P et al (2008) Cost effectiveness of enzyme immunoassay and immunoblot testing for the diagnosis of syphilis. Int J STD AIDS 19:393–399
30. Owasu-Edusei K, Peterman TA, Ballard RC (2011) Serologic testing of syphilis in the United Sates: a cost-effective analysis of two screening algorithm. Sex Transm Dis 38(1):1–7
31. Merritt HH, Adams RD, Solomon HC (1946) Neurosyphilis, Oxford University Press, New York
32. Stokes JH, Beerman H, Ingraham NR (1944) Modern clinical syphilology: diagnosis, treatment, case study, 3rd edn. WB Saunders, Philadelphia
33. Taylor MM, Aynalem G, Olea LM, He P, Smith LV, Kerndt PR (2008) A consequence of the syphilis epidemic among men who have sex with men (MSM): neurosyphilis in Los Angeles, 2001–2004. Sex Transm Dis 35(5):430
34. Walker DG, Walker GJA (2002) Forgotten but not gone: the continuing scourge of congenital syphilis. Lancet Infect Dis 2:432–436
35. Walker DG, Walker GJ (2004) Prevention of congenital syphilis. Time of action. Bull World Health Organ 82(6):401

36. Jones H, Taylor D, Montgomery CA, Patrick DH, Money D, Vipond JC, Morshed MG, Ruissard DA, Rekart ML (2005) Prenatal and congenital syphilis in British Columbia. J Obstet Gynaecol Can 27(5):467–472
37. Dean GL (2006) Near-patient testing will not improve the control of sexually transmitted infections. Sex Transm Dis 82:509–512
38. Tucker JD, Brown LB, Yin Y-P, Chen X-S, Cohen MS (2010) Accelerating worldwide syphilis screening through rapid testing: a systematic review. Lancet Infect Dis 10:381–386

# Rapid Spreading and Immune Evasion by Vaccinia Virus

**Geoffrey L. Smith**

**Abstract** Vaccinia virus (VACV) is the live vaccine that was used to eradicate smallpox, a feat achieved in 1977 and certified by the World Health Organization in 1980. Since 1980, research with VACV has continued in part because of the development of techniques to genetically manipulate VACV and create live VACV strains expressing foreign genes. These recombinant VACVs can be used as live vaccines against other infectious diseases and cancers, and as a powerful tool to study virus pathogenesis, immunology, cell biology, and virus–host interactions. This short article describes two examples of how enduring interest in VACV has revealed new features of VACV biology and the immune system.

**Keywords** Vaccinia virus · Smallpox vaccine · Immune evasion · Virus spreading mechanism

## Introduction

In 1796 Jenner vaccinated a 9-year-old boy, James Phipps, with cowpox virus taken from the hand of milkmaid, Sarah Nelmes, and showed by challenging the boy subsequently with variola virus, the etiological agent of smallpox, that he was protected against smallpox [1]. Just 5 years later, Jenner predicted that the "benefits of this practice were too manifest to admit of controversy and that the annihilation of smallpox, the most dreadful scourge of the human species, must be the final result of this practice" [2]. This prophecy was proved correct and the eradication of smallpox was certified by the World Health Organization (WHO) in 1980. Smallpox remains the first and only human infectious disease to be eradicated [3].

G. L. Smith (✉)
Department of Pathology, University of Cambridge, Tennis Court Road, Cambridge CB2 1QP, UK
e-mail: gls37@cam.ac.uk

R. Adhikari and S. Thapa (eds.), *Infectious Diseases and Nanomedicine II*, Advances in Experimental Medicine and Biology 808,
DOI: 10.1007/978-81-322-1774-9_6, 

Despite this success, the vaccine used to eradicate smallpox has an unknown origin and host and remains an enigma of virology [4]. Although cowpox virus was believed to have been used by Jenner, the vaccine used in the modern era is a distinct species of orthopoxvirus of unknown origin that became known as VACV after the Latin word vacca for cow. In 1939, Alan Downie showed that all the smallpox vaccine strains in use in the twentieth century shared common biological properties that were distinct from cowpox virus [5]. Although VACV and cowpox virus are distinct, they belong to the same genus (*Orthopoxvirus*) as variola virus and so each provide immunological protection against smallpox.

Scarcely had the ink dried on the WHO certificate proclaiming smallpox eradication, than two research groups in USA developed techniques to genetically manipulate VACV and express foreign proteins from live VACV strains [6, 7]. These recombinant viruses showed potential as live vaccines against the organism from which the other protein was derived [8–10]. Such viruses also proved valuable research tools to evaluate the immune response to virus infections [11–13] and to express virtually any gene in a wide range of mammalian cells [14]. Although VACV was used to achieve the global eradication of a disease, it had an imperfect safety record compared to other vaccines [15] and so if VACV was to be reused as a vaccine against other diseases it was recognized that its safety needed to be improved. To improve the safety of VACV as a human vaccine, research groups set out to understand VACV virulence and, in particular, how it interacts with the host to spread rapidly and to suppress the innate immune system. It was this research that led to the two projects described below.

## VACV Spread

The replication of VACV in an infected cell produces multiple distinct forms of infectious virus particle, for reviews see [16–18]. Replication takes place in cytoplasmic factories and the first type of virion produced is called intracellular mature virus (IMV) and is surrounded by a single lipid envelop [19, 20]. This form represents the majority of progeny virus and is very stable when freeze dried. Some IMV particles are transported on microtubules from the virus factory in the cytoplasm to a site near the microtubule organizing center [21] where the virion becomes wrapped by a double layer of host membrane that has been modified by the incorporation of virus proteins [22, 23]. The resultant virion is surrounded by three membranes and is called intracellular enveloped virus (IEV). IEV particles are transported on microtubules to the cell surface [24–27] where the outer membrane fuses with the plasma membrane to expose a virion on the cell surface by exocytosis. This double enveloped virus particle is called cell-associated enveloped virus (CEV) and functions to induce the polymerization of actin from beneath the plasma membrane where the virion is attached [28–31]. This growing actin tail then propels CEV particles toward uninfected cells to initiate a new replication cycle. Finally, some double enveloped virions are released from the cell, rather than being retained on the surface, and are called extracellular enveloped virus (EEV) [32]. The four types of virion have differing functions. The IMV particle is very stable and is released from infected cells only following cell lysis. Its stability makes it effective to transmit infection between hosts, but it poorly adapted to spread infection within a host because of its sensitivity to destruction by complement and ease of neutralization by antibodies. The IEV particle represents a transient virion that exploits microtubule-based transport to move the unit of infectivity from the cell center to the periphery. The CEV particle functions to induce the formation of actin tails from the cell surface to spread virus from cell to cell, and EEV is important for the longer range spread of virus in cell culture or in vivo [32].

VACV is easy to grow in cell culture and replicates in the great majority of cultured cell types. It forms large round plaques in 2–3 days in cell monolayers with a solid or semi-solid overlay. But if the infected cell monolayer is cultured in liquid overlay, the plaques appear as comets, with the head of the comet representing the primary plaque and the comet tail representing secondary infections caused by release of EEV from the primary plaque [33, 34]. VACV strains that release larger amounts of EEV form more pronounced comets [35]. The formation of comets is inhibited either by the solid overlay to prevent virus diffusion, or by antibodies to EEV surface proteins [32–34].

Ever since the development of the plaque assay for animal viruses [36] it had been thought that plaque formation was an iterative process in which a virus particle infects a cell, replicates, and then is released to infect surrounding cells. The virus then replicates in these new cells and is again released to spread the infection further. If this hypothesis was correct and plaque formation represents an iterative process of infection, release, and reinfection, then the rate at which a plaque formed should be limited by the replication kinetics of the virus in each infected cell. However, up to 2010, studies investigating whether the size of the plaque formed by different viruses matched their replication kinetics were lacking.

To study VACV plaque formation, we measured the size of plaques formed on monolayers of BSC-1 cells and used phase contrast microscopy to determine the average distance between adjacent cell nuclei and thereby the average cell size [37]. In this way, the number of cells that the virus had destroyed to form a plaque over 3 days could be determined. By relating the time taken to form a plaque to the number of cells within the radius of a plaque, the rate of virus spread (hours per cell) was calculated. This revealed that VACV was spreading across each cell in less than 2 h [37]. This was extraordinary, given that the replication of the virus in each cell was known to take a minimum of 5 h, and the majority of infectious virus was not produced until later than that. Measurements of plaque formation by live video microscopy confirmed this rapid spread and refined the rate of spread to one cell every 1.2 h [37]. These observations indicated that VACV plaque formation was not just a simple case of iterative replication, release, and reinfection but that there must be another mechanism.

To address how VACV was spreading so fast a recombinant VACV engineered to express the enhanced green fluorescent protein (EGFP) fused to the late virus

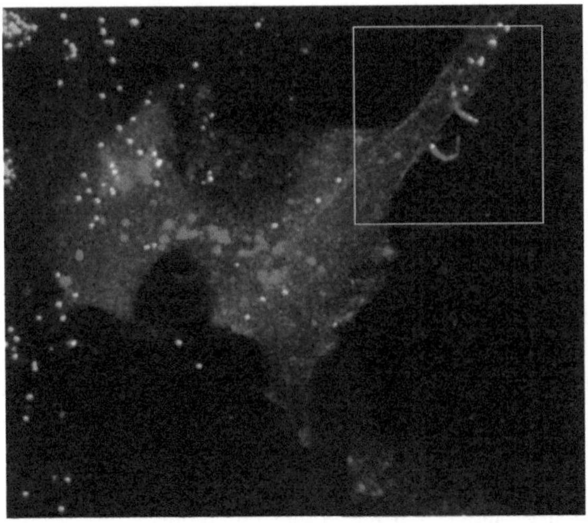

**Fig. 1** The spread of VACV is enhanced by the repulsion of superinfecting virions on actin tails from infected cells. The image shows a cell that has been engineered to express cherry fluorescent protein fused to actin and so this cell is *red*. Adjacent to this cell, to the left, there is a normal cell that has been infected with a VACV strain expressing the green fluorescent protein fused to a VACV capsid protein that is expressed late during infection. Factories where new VACV particles are being formed are evident as aggregates of *green dots* (such as in the extreme *top leftcorner*) and individual virus particles are evident as individual *green dots*. The *redcell*, although infected, has not produced new virus particles because it has not gone green. Yet, on the surface of the *redcell* (within the *whitebox*) there are green virus particles being driven by a red actin tail. This image is one still shot taken from a video showing the movement of these green virus particles [37]

capsid protein A4 [38] was utilized. Cells infected by this virus become green when the virus enters the late phase of virus gene expression, and green virus particles can be seen forming in virus factories and emanating from these green factories to the cell surface and hence to uninfected cells (Fig. 1).

Examination of the edge of a plaque formed by the EGFP—expressing VACV showed that there were green virus particles on the surface of cells some distance from the nearest green cell from which the virus must have originated [37] (Fig. 2). Furthermore, by staining these cells with phalloidin to detect actin, it was evident that some of these green virus particles were at the tip of an actin tail. Since the cells on which these green virions were attached to an actin tail were not green, the virus must have come from a different cell. This suggested that the virus particle and actin tail derived from different cells.

To examine this more closely we created a cell line in which actin was fused with the cherry fluorescent protein. By mixing these cells with normal cells, a monolayer of cells was created that represented a mosaic of red and normal cells. This monolayer was then infected by a few green virus particles and left until small plaques started to form. The cells adjacent to the edge of a green plaque were then examined carefully by microscopy. Such analysis identified red cells, which

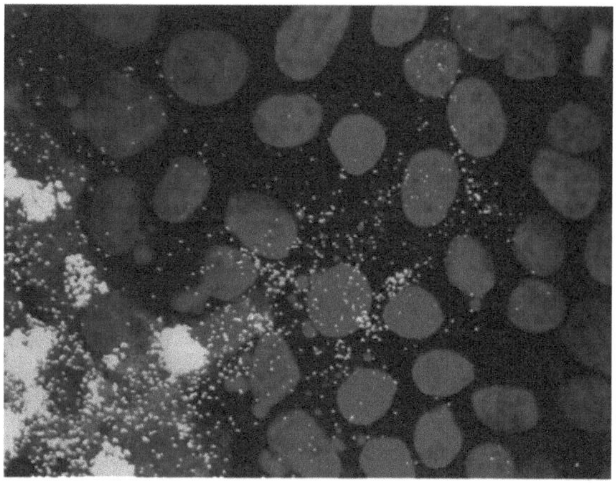

**Fig. 2** Infection of cells by VACV produces cytoplasmic factories containing virus DNA. The image shows a monolayer of BSC-1 cells that has been infected for 1 day with a few infectious particles of a strain of VACV expressing the *green*fluorescent protein (GFP) fused to a VACV capsid protein. The image shows the edge of a plaque formed by VACV on the cell monolayer with the infected cells showing *green* due to expression of the GFP. The cells are also stained with DAPI, to detect DNA. The cell nuclei and the cytoplasmic virus factories of infected cells containing virus DNA are *blue*. It is evident that infected cells contained large aggregates of cytoplasmic DNA. This PAMP would be detected by host cell PRRs, leading to activation of the innate immune response, unless VACV employed strategies to inhibit this process, such as protein C16 [53]

had not gone green, but which had on their surface green virus particles driven by a red actin tail (Fig. 1). This proved that the actin tail and the green virus particle had derived from different cells and, therefore, that somehow VACV was inducing actin tails on the surface of cells from which virions had not been formed [37].

To address the mechanism by which this occurred, we hypothesized that early after infection, VACV was expressing one or more proteins on the cell surface that marked the cell as infected. Subsequently, when an additional virus particle tried to infect that cell (superinfection), it encountered the virus surface protein(s) and this interaction triggered the formation of actin tails to drive the superinfecting virion away to uninfected cells. In considering what type of protein(s) might be responsible, it seemed that such molecule(s) would need to be (1) expressed early after infection, (2) be on the cell surface, and (3) be involved in the induction of actin tails. An examination of the literature suggested three proteins fitted these criteria, called A33, A36, and B5. A curious feature of each of these proteins was that they were expressed both early and late during infection, despite the only known functions of these proteins being later during infection [39–41]. To address if the early expression of these proteins was necessary for rapid virus spread, recombinant VACVs were engineered in which each gene was expressed only from a late virus promoter and the size of plaque formed by such viruses was compared to that formed by wild-type virus and recombinant viruses lacking each gene separately. Such comparison showed that viruses expressing the A33 and

A36 proteins only late during infection formed small plaques, intermediate in size between those formed by wild-type and deletion mutant. In contrast, expression of the B5 protein only late during infection caused no alteration in plaque size [37].

The A33 and A36 proteins were known to be expressed on the cell surface and to form a complex [42–47]. A33 is a type II membrane glycoprotein with the majority of the polypeptide exposed outside the cell [43]. A36, in contrast, is a type IB membrane protein with a single transmembrane region and the majority of the protein within the cytosol [40, 46]. Both proteins are necessary for the formation of actin tails from the surface of cells producing new virus particles [45–47]. To determine if these proteins were sufficient for the formation of actin tails, cell lines were constructed using lentivirus vectors, in which either one or both proteins were expressed. EEV particles were then added to the surface of these cells and the cells were examined for actin tail formation. Remarkably, addition of EEV induced actin tail formation from cells expressing both proteins within 15 min. Either protein alone was insufficient and only EEV, not IMV or herpes simplex virus (HSV), was able to induce actin tails [37].

In summary, VACV has evolved a mechanism to induce the rapid spread of virus without the need to replicate in each cell as it spreads. In the first cell infected, the virus undergoes its complete replication cycle and new virus particles are pushed on actin tails away from the infected cell to infect the adjacent cell. In this cell, the A33 and A36 proteins are expressed early after infection and are transported to the cell surface as a complex. On the cell surface they mark the cell as infected and can interact with additional EEV or CEV particles trying to superinfect the same cell. Such interaction induces actin polymerization beneath the plasma membrane where the superinfecting virus particle has bound, and this actin polymerization drives the particle away toward additional cells. Thus, superinfecting virions are repelled and bounced or surfed across the surface of infected cells until they reach an uninfected cell [37].

This mechanism for rapid spread is logical and Darwinian. It makes little sense for a virus to superinfect the same cell it has infected already, but a lot of sense to spread rapidly to uninfected cells before the host can mount an immune response to infection. It is likely that other viruses may have evolved mechanisms for rapid spread and for HSV there is evidence that the plaques formed are also too large to be explained by the known replication kinetics of the virus [37].

## Evasion of Innate Immunity

Even if a virus can spread rapidly between cells, the infection will be sensed by host pattern recognition receptors (PRRs) either on the surface of cells or within infected cells. Such recognition will lead to activation of innate immune signaling pathways and production of interferons (IFNs) and pro-inflammatory cytokines and chemokines [48, 49]. For many viruses, it is their nucleic acid that represents the pathogen-associated molecular pattern (PAMP) that is detected by the PRRs.

The virus nucleic acid is distinguished from host nucleic acid by virtue of either being structurally different from that of the host, or being present in the wrong subcellular compartment; for instance, DNA in the cytoplasm would be highly unusual for normal cells. PRRs for nucleic acids exist either in the membrane of endosomes, such as toll-like receptors 3 and 9, or within the cytosol. Cytosolic PRRs for foreign RNA have been known for some time, such as RIG-I or MDA-5, but those for DNA have been discovered more recently and remain less well characterized [50]. Given that VACV replicates its large dsDNA genome in the cell cytoplasm and can produce up to 10,000 copies of the genome within a few hours [51], it is highly likely that this DNA will be sensed by host PRRs to initiate an innate immune response (Fig. 2). The pressure that this exerts on VACV will have driven the evolution of countermeasures. The following project illustrates how VACV protein C16 [52] inhibits the innate response to virus DNA [53] and how the study of this protein led to the discovery of a new PPR for DNA called DNA protein kinase (DNA-PK) [54].

VACV strain Western Reserve (WR) protein C16 is expressed early during infection and is reasonably well conserved between different VACV strains and other orthopoxviruses and chordopoxviruses. Nonetheless, the isolation of a VACV WR viable deletion mutant lacking the *C16L* gene showed that C16 was nonessential for VACV replication in cell culture [52]. This VACV strain lacking gene *C16L*, called vΔC16, replicated and spread normally in cell culture, but was attenuated compared to control parental and revertant viruses when the virus virulence was analyzed by intradermal infection of mice [52]. Characterization of the innate immune response showed that infection by vΔ16 induced a greater influx of inflammatory cells than control viruses, indicating that the protein was somehow blocking the innate response to infection [52].

To address the mechanism by which C16 contributed to virus virulence, the C16 protein was tagged with a tandem affinity purification (TAP) tag [55] and expressed in cells. After purification of the C16 protein via its TAP tag, the binding partners were identified by SDS–polyacrylamide gel electrophoresis (PAGE) and mass spectrometry. This unambiguously identified proteins Ku70 and Ku86 as C16 binding partners [53]. The Ku proteins form a heterodimer and associate with DNA–protein kinase (DNA-PKcs) to form a trimeric complex called DNA-PK [56–58]. DNA-PK is an abundant and well-characterized complex and has a critical role in development of adaptive immunity due to its role in nonhomologous DNA joining that is needed for VDJ recombination during the development of antibody and T cell receptor diversity. However, a role for DNA-PK in the cytoplasm was less clear.

Mechanistically, it was shown that C16 binding to the Ku heterodimer prevents the interaction of Ku with DNA and this restricts the recruitment of DNA-PKcs to the DNA–Ku complex [53]. But the question remained as to why VACV should have evolved a protein to block recruitment of DNA-PK components to DNA and how this linked to the in vivo phenotype of vΔC16 that was characterized by a stronger innate immune response the infection. This prompted an investigation of the role of DNA-PK in innate immunity and led to the conclusion that DNA-PK is a PRR for cytoplasmic DNA and this interaction leads to an IRF3-dependent innate immune response [54].

To identify intracellular PRRs for foreign DNA, a short double-stranded immunostimulatory DNA molecule was tagged with biotin and transfected into human embryonic kidney (HEK) 293 cells or murine embryonic fibroblasts (MEFs). One hour later, the cells were separated into cytoplasmic and nuclear fractions and the cytoplasmic DNA was affinity purified via its biotin tag and associated proteins were detected by SDS-PAGE and mass spectrometry. This identified DNA-PKcs, Ku70, and Ku86 as abundant DNA binding proteins in the cytoplasm. In parallel, it was shown that transfection of such DNA into MEFs induced an innate immune response as illustrated by the upregulation of the chemokine CXCL10, interleukin (IL)-6, and IFN$\beta$ [54]. Furthermore, similar immune responses were induced by DNA from *E. coli* or VACV and by the synthetic DNA poly (dA:dT). To determine if components of the DNA-PK complex were necessary for this innate immune response, primary MEFs lacking DNA-PKcs, or MEFs lacking either Ku subunit, were tested and found to be defective in responding to transfected DNA. The defect in DNA-PKcs -/- MEFs was found to be more profound than that in Ku86 -/- MEFs, consistent with the prior observation that DNA-PKcs could still bind DNA (albeit weakly) in the absence of Ku.

The defect in innate immune signaling was specific to DNA, because the responses to RNA or LPS were normal in cells lacking DNA-PKcs. Further analysis showed that the signaling pathway activated by DNA-PK binding to DNA required TANK-binding kinase 1 (TBK-1), STING and interferon regulatory factor (IRF)3, but was independent of toll-like receptor signaling and other innate immune signaling pathways. The defect in responding to DNA was also observed following infection with a DNA virus but not an RNA virus, and, importantly, was also manifest in vivo following transfection of DNA or infection with a DNA virus in DNA-PKcs -/- mice. In contrast, the response to RNA or RNA virus infection was normal in cultured cells or mice lacking DNA-PKcs.

This study established that DNA-PK was a PRR for DNA and induced an innate immune response via the TBK-1, STING, and IRF3 signaling pathway. This provided a plausible explanation for why VACV had evolved a protein, C16, to bind to the Ku complex and suggested that C16 might inhibit innate immune signaling. This hypothesis was tested by introducing a plasmid encoding C16 into cells and showing that the innate response to DNA was inhibited. If the plasmid was digested with a restriction enzyme that disrupted the C16 open reading frame (ORF), the inhibition of DNA signaling was lost, but if the plasmid was digested with an enzyme that cut elsewhere in the plasmid and did not disrupt C16 expression, the inhibition was retained. Mechanistically, it was found that C16 did not bind to DNA itself, or to DNA-PKcs, but was able to prevent the recruitment of these components to DNA. Lastly, the relevance of C16 and DNA-PK to innate immunity was demonstrated in vivo by the observation that infection with v$\Delta$C16 produced a greater inflammatory response than infection with control viruses expressing C16.

In summary, VACV protein C16 binds the Ku complex and thereby inhibits innate immune signaling triggered by recognition of cytoplasmic DNA by DNA-PK. The interaction of C16 with Ku70/Ku86 prompted investigation of DNA-PK in innate immune signaling and led to the discovery that DNA-PK functions as a PRR in innate immunity. Thus DNA-PK has important roles in both innate and adaptive immunity and DNA repair.

## Concluding Remarks

These two projects illustrate how the study of VACV continues to lead to enhanced understanding of basic biological principles. In the first example, the rate of spread of VACV was shown to be faster than predicted from its known replication kinetics and a mechanism for this was elucidated. The repulsion of super-infecting virions from infected cells accelerated virus spread and enabled virus particles to travel across several cells without the need to replicate in each cell along the way. This established that virus spread is not simply replication, release, and reinfection, but rather is rapid spread without replication in every cell. The discovery of this mechanism and the proteins involved provides a target for development of antiviral drugs. It also suggests that other viruses too might have evolved mechanisms for rapid spread.

The second example illustrates how VACV has evolved a mechanism to block activation of the innate immune system that is triggered by recognition of cytoplasmic DNA. This led to the discovery of DNA-PK as a new PRR for innate immunity. The importance of DNA-PK as a PRR in vivo was demonstrated first by the defect of DNA-PKcs -/- mice in responding to DNA or DNA virus infection, and second by the fact that VACV has evolved a strategy to block the function of DNA-PK as a PRR to escape IRF3-dependent innate immunity.

**Acknowledgments** Work in my laboratory has been supported generously by grants from the UK Medical Research Council and the Wellcome Trust, and I thank these funding agencies most warmly for this support. GLS is a Wellcome Trust Principal Research Fellow. I am also deeply grateful to the Nepalese Association of Medical Microbiology for the kind invitation to visit Nepal and speak at this First International Conference on Infectious Diseases and Nanomedicine-2012 (ICIDN-2012). On a personal note, it was a special honor to visit the country of the Gurkhas who were comrades in arms with my grandfather Professor Sir Ralph L Turner MC during the Great War and who saved his life in 1917, in Palestine. In the preface to R.L. Turner's dictionary of the Nepali language [59] he expressed his lifelong respect and affection for the Gurkha soldier by words that are now engraved on the British Memorial to the Gurkha soldier in Whitehall, London and were unveiled by Her Majesty Queen Elizabeth II in 1997: "*Bravest of the brave, most generous of the generous, never had country more faithful friends than you*". To be able to recite these words before the people and President of Nepal was an unforgettable experience.

## References

1. Jenner E (1798) An enquiry into the causes and effects of variolae vaccinae, a disease discovered in some Western Countries of England, particularly Gloucestershire, and known by the name of cowpox. Cassell, London Reprinted by Cassell, 1896
2. Jenner E (1801) The origin of the vaccine inoculation. D.N. Shury, London
3. Fenner F, Anderson DA, Arita I, Jezek Z, Ladnyi ID (1988) Smallpox and its eradication. World Health Organisation, Geneva
4. Baxby D (1981) Jenner's smallpox vaccine: the riddle of the origin of vaccinia virus. Heinemann, London
5. Downie AW (1939) Immunological relationship of the virus of spontaneous cowpox to vaccinia virus. Br J Exp Pathol 20:158–176
6. Panicali D, Paoletti E (1982) Construction of poxviruses as cloning vectors: insertion of the thymidine kinase gene from herpes simplex virus into the DNA of infectious vaccinia virus. Proc Natl Acad Sci USA 79:4927–4931
7. Mackett M, Smith GL, Moss B (1982) Vaccinia virus: a selectable eukaryotic cloning and expression vector. Proc Natl Acad Sci USA 79:7415–7419
8. Smith GL, Mackett M, Moss B (1983) Infectious vaccinia virus recombinants that express hepatitis B virus surface antigen. Nature 302:490–495
9. Smith GL, Murphy BR, Moss B (1983) Construction and characterization of an infectious vaccinia virus recombinant that expresses the influenza hemagglutinin gene and induces resistance to influenza virus infection in hamsters. Proc Natl Acad Sci USA 80:7155–7159
10. Panicali D, Davis SW, Weinberg RL, Paoletti E (1983) Construction of live vaccines by using genetically engineered poxviruses: biological activity of recombinant vaccinia virus expressing influenza virus hemagglutinin. Proc Natl Acad Sci USA 80:5364–5368
11. Bennink JR, Yewdell JW, Smith GL, Moss B (1986) Recognition of cloned influenza virus hemagglutinin gene products by cytotoxic T lymphocytes. J Virol 57:786–791
12. Osman M, Kubo T, Gill J, Neipel F, Becker M et al (1999) Identification of human herpesvirus 8-specific cytotoxic T-cell responses. J Virol 73:6136–6140
13. Yewdell JW, Bennink JR, Smith GL, Moss B (1985) Influenza A virus nucleoprotein is a major target antigen for cross-reactive anti-influenza A virus cytotoxic T lymphocytes. Proc Natl Acad Sci USA 82:1785–1789
14. Mackett M, Smith GL (1986) Vaccinia virus expression vectors. J Gen Virol 67:2067–2082
15. Lane JM, Ruben FL, Neff JM, Millar JD (1969) Complications of smallpox vaccination, 1968: national surveillance in the United States. N Engl J Med 281:1201–1208
16. Roberts KL, Smith GL (2008) Vaccinia virus morphogenesis and dissemination. Trends Microbiol 16:472–479
17. Smith GL, Vanderplasschen A, Law M (2002) The formation and function of extracellular enveloped vaccinia virus. J Gen Virol 83:2915–2931
18. Condit RC, Moussatche N, Traktman P (2006) In a nutshell: structure and assembly of the vaccinia virion. Adv Virus Res 66:31–124
19. Hollinshead M, Vanderplasschen A, Smith GL, Vaux DJ (1999) Vaccinia virus intracellular mature virions contain only one lipid membrane. J Virol 73:1503–1517
20. Dales S, Siminovitch L (1961) The development of vaccinia virus in Earle's L strain cells as examined by electron microscopy. J Biophys Biochem Cytol 10:475–503
21. Sanderson CM, Hollinshead M, Smith GL (2000) The vaccinia virus A27L protein is needed for the microtubule-dependent transport of intracellular mature virus particles. J Gen Virol 81:47–58
22. Schmelz M, Sodeik B, Ericsson M, Wolffe EJ, Shida H et al (1994) Assembly of vaccinia virus: the second wrapping cisterna is derived from the trans Golgi network. J Virol 68:130–147

23. Tooze J, Hollinshead M, Reis B, Radsak K, Kern H (1993) Progeny vaccinia and human cytomegalovirus particles utilize early endosomal cisternae for their envelopes. Eur J Cell Biol 60:163–178
24. Hollinshead M, Rodger G, Van Eijl H, Law M, Hollinshead R et al (2001) Vaccinia virus utilizes microtubules for movement to the cell surface. J Cell Biol 154:389–402
25. Ward BM, Moss B (2001) Vaccinia virus intracellular movement is associated with microtubules and independent of actin tails. J Virol 75:11651–11663
26. Rietdorf J, Ploubidou A, Reckmann I, Holmstrom A, Frischknecht F et al (2001) Kinesin-dependent movement on microtubules precedes actin-based motility of vaccinia virus. Nat Cell Biol 3:992–1000
27. Geada MM, Galindo I, Lorenzo MM, Perdiguero B, Blasco R (2001) Movements of vaccinia virus intracellular enveloped virions with GFP tagged to the F13L envelope protein. J Gen Virol 82:2747–2760
28. Cudmore S, Cossart P, Griffiths G, Way M (1995) Actin-based motility of vaccinia virus. Nature 378:636–638
29. Stokes GV (1976) High-voltage electron microscope study of the release of vaccinia virus from whole cells. J Virol 18:636–643
30. Hiller G, Weber K, Schneider L, Parajsz C, Jungwirth C (1979) Interaction of assembled progeny pox viruses with the cellular cytoskeleton. Virology 98:142–153
31. Blasco R, Cole NB, Moss B (1991) Sequence analysis, expression, and deletion of a vaccinia virus gene encoding a homolog of profilin, a eukaryotic actin-binding protein. J Virol 65:4598–4608
32. Payne LG (1980) Significance of extracellular enveloped virus in the in vitro and in vivo dissemination of vaccinia. J Gen Virol 50:89–100
33. Appleyard G, Hapel AJ, Boulter EA (1971) An antigenic difference between intracellular and extracellular rabbitpox virus. J Gen Virol 13:9–17
34. Law M, Hollinshead R, Smith GL (2002) Antibody-sensitive and antibody-resistant cell-to-cell spread by vaccinia virus: role of the A33R protein in antibody-resistant spread. J Gen Virol 83:209–222
35. Payne LG (1979) Identification of the vaccinia hemagglutinin polypeptide from a cell system yielding large amounts of extracellular enveloped virus. J Virol 31:147–155
36. Dulbecco R (1952) Production of plaques in monolayer tissue cultures by single particles of an animal virus. Proc Natl Acad Sci USA 38:747–752
37. Doceul V, Hollinshead M, van der Linden L, Smith GL (2010) Repulsion of superinfecting virions: a mechanism for rapid virus spread. Science 327:873–876
38. Carter GC, Rodger G, Murphy BJ, Law M, Krauss O et al (2003) Vaccinia virus cores are transported on microtubules. J Gen Virol 84:2443–2458
39. Engelstad M, Howard ST, Smith GL (1992) A constitutively expressed vaccinia gene encodes a 42-kDa glycoprotein related to complement control factors that forms part of the extracellular virus envelope. Virology 188:801–810
40. Parkinson JE, Smith GL (1994) Vaccinia virus gene A36R encodes a M(r) 43–50 K protein on the surface of extracellular enveloped virus. Virology 204:376–390
41. Roper RL, Wolffe EJ, Weisberg A, Moss B (1998) The envelope protein encoded by the A33R gene is required for formation of actin-containing microvilli and efficient cell-to-cell spread of vaccinia virus. J Virol 72:4192–4204
42. Lorenzo MM, Galindo I, Griffiths G, Blasco R (2000) Intracellular localization of vaccinia virus extracellular enveloped virus envelope proteins individually expressed using a Semliki Forest virus replicon. J Virol 74:10535–10550
43. Roper RL, Payne LG, Moss B (1996) Extracellular vaccinia virus envelope glycoprotein encoded by the A33R gene. J Virol 70:3753–3762
44. Rottger S, Frischknecht F, Reckmann I, Smith GL, Way M (1999) Interactions between vaccinia virus IEV membrane proteins and their roles in IEV assembly and actin tail formation. J Virol 73:2863–2875

45. Sanderson CM, Frischknecht F, Way M, Hollinshead M, Smith GL (1998) Roles of vaccinia virus EEV-specific proteins in intracellular actin tail formation and low pH-induced cell–cell fusion. J Gen Virol 79:1415–1425
46. van Eijl H, Hollinshead M, Smith GL (2000) The vaccinia virus A36R protein is a type Ib membrane protein present on intracellular but not extracellular enveloped virus particles. Virology 271:26–36
47. Wolffe EJ, Weisberg AS, Moss B (2001) The vaccinia virus A33R protein provides a chaperone function for viral membrane localization and tyrosine phosphorylation of the A36R protein. J Virol 75:303–310
48. Akira S, Uematsu S, Takeuchi O (2006) Pathogen recognition and innate immunity. Cell 124:783–801
49. Medzhitov R (2007) Recognition of microorganisms and activation of the immune response. Nature 449:819–826
50. Paludan SR, Bowie AG (2013) Immune sensing of DNA. Immunity 38:855–869
51. Moss B (2007) Poxviridae: the viruses and their replicaton. In: Knipe DM (ed) Fields virology, 5th edn. Lippincott Williams & Wilkins, Philadelphia, pp 2905–2946
52. Fahy AS, Clark RH, Glyde EF, Smith GL (2008) Vaccinia virus protein C16 acts intracellularly to modulate the host response and promote virulence. J Gen Virol 89:2377–2387
53. Peters NE, Ferguson BJ, Mazzon M, Fahy AS, Krysztofinska E et al (2013) A mechanism for the inhibition of DNA-PK-mediated DNA sensing by vaccinia virus. PLoS Pathog 9: e1003649
54. Ferguson BJ, Mansur DS, Peters NE, Ren H, Smith GL (2012) DNA-PK is a DNA sensor for IRF-3-dependent innate immunity. elife 1: e00047
55. Gloeckner CJ, Boldt K, Schumacher A, Roepman R, Ueffing M (2007) A novel tandem affinity purification strategy for the efficient isolation and characterisation of native protein complexes. Proteomics 7:4228–4234
56. Collis SJ, DeWeese TL, Jeggo PA, Parker AR (2005) The life and death of DNA-PK. Oncogene 24:949–961
57. Spagnolo L, Rivera-Calzada A, Pearl LH, Llorca O (2006) Three-dimensional structure of the human DNA-PKcs/Ku70/Ku80 complex assembled on DNA and its implications for DNA DSB repair. Mol Cell 22:511–519
58. Rivera-Calzada A, Spagnolo L, Pearl LH, Llorca O (2007) Structural model of full-length human Ku70-Ku80 heterodimer and its recognition of DNA and DNA-PKcs. EMBO Rep 8:56–62
59. Turner RL (1931) A comparative and etymological dictionary of the Nepali language. Oxford University Press, Oxford

# Nanonization Increases the Antileishmanial Efficacy of Amphotericin B: An Ex Vivo Approach

**Krishna Das Manandhar, Thakur Prasad Yadav, Vijay Kumar Prajapati, Om Basukala, Ram Prasad Aganja, Anuradha Dude, Onkar Nath Shrivastav and Shyam Sundar**

**Abstract** With widespread resistance to pentavalent antimonial in the endemic eastern terai belt of Nepal and Bihar, India, Amphotericin B deoxycholate is now the first-line antileishmanial drug for the treatment of visceral leishmaniasis (VL). However, universal occurrence of infusion-related fever and rigors with amphotericin B (AmB), occasional serious life-threatening toxicities like cardiotoxicity, anaphylaxis, hypokalemia, and nephrotoxicity are major barriers to its use in areas with limited medical facilities. Liposomal amphotericins, however, are devoid of adverse effects, high cost makes it unaffordable. We had formulated nanoparticles (10–20 nm) from amphotericin B deoxycholate (1–2 μm) applying high pressure (150 atm) milling homogenization in argon atmosphere and tested its ex vivo efficacy in *Leishmania* infected J774A cell line and peritoneal macrophage. The ex vivo $ED_{50}$ for intracellular amastigotes in peritoneal macrophage by nano-amphotericin was $0.0027 \pm 0.001$ μg/mL which was significantly less ($p = 0.0029$) than the required dose of amphotericin B ($0.0426 \pm 0.003$ μg/mL). Similarly, in J774A cell line, 50 % of intracellular amastigotes were cleared by

K. D. Manandhar (✉) · O. Basukala · R. P. Aganja
Central Department of Biotechnology, Institute of Science and Technology, Tribhuvan University, Kathmandu, Nepal
e-mail: krishna.manandhar@gmail.com

K. D. Manandhar · V. K. Prajapati · S. Sundar
Infectious Disease Research Laboratory, Department of Medicine, Institute of Medical Science, Banaras Hindu University, Varanasi, India

T. P. Yadav · O. N. Shrivastav
Department of Physics, Banaras Hindu University, Varanasi, India

V. K. Prajapati
Department of Biochemistry, The Central University of Rajasthan, Ajmer, Rajasthan, India

O. Basukala
Nepal Academy of Science and Technology (NAST), Khumaltar, Lalitpur, Nepal

A. Dude
Central Drug Research Institute, Lucknow, India

R. Adhikari and S. Thapa (eds.), *Infectious Diseases and Nanomedicine II*, Advances in Experimental Medicine and Biology 808,
DOI: 10.1007/978-81-322-1774-9_7, 

0.0038 ± 0.001 μg/mL of nano-amphotericin while the dose was a bit more for AmB (0.0196 ± 0.001 μg/mL) illustrating the significant difference (*p* value, 0.0122). The nanoformulation has also shown high efficacy (ED50, 0.0028–0.0035 μg/mL) in inhibition of infected macrophage count. The new formulation accumulated to spleen, the targeted organ, 7 days after inoculation of drug to the infected hamster as traced in vivo by TEM convincing it as potential drug. Given a favorable safety profile and very low cost of production contemplated, it may prove to be a feasible alternative for conventional amphotericin B.

**Keywords** Amphotericin B · Visceral leishmaniasis · Leishmania donovani · Nanoparticles · Hamster · Ex vivo

## Visceral Leishmaniasis

Visceral leishmaniasis (VL/kala-azar), caused by *L. donovani* in Indian subcontinent, is the most severe form of infectious disease which unless treated is uniformly fatal. The disease claims for 75,000 deaths per anum [1, 2] and is most prevalent in the poorest of the poor community. Nepal, India, Bangladesh, Sudan, and Brazil account about 90 % of the world's VL burden [3, 4]. Transmission of VL in the Indian subcontinent is anthroponotic, and female sandfly, *Phlebotomus argentipes* is the vector. This disease is characterized by fever for more than 2 weeks with rigor and chills, hepatosplenomegaly, lymphadenopathy, pancytopenia, hypergammaglobulinaemia, hypoalbuminemia, and cachexia [5].

### *Treatment Practices and Lacunae*

It is a completely fatal tropical disease if not treated. The treatment options for VL are limited and unsatisfactory as the drugs available are mostly parenteral with serious toxicity. Besides, widespread nonresponsiveness to pentavalent antimonial have been continued [6, 7]. Amphotericin B (AmB) is being used as first-line drug in the Indian subcontinent as it cures nearly 100 % though adverse drug reactions have been the real bottleneck. Oral miltefosine and paromomycin are other new available options along with other drugs, nevertheless, due to treatment failures, nonavailability, or contraindications, amphotericin B remains the most important drug. A total dose of 15–20 mg/kg [8] intravenous infusion has to be given on alternate days necessitating prolonged hospitalization, which leads to denial of treatment to a large proportion of patients because of lack of hospital beds, expertise, etc. In addition, occasional serious adverse effect entails adequate monitoring in patients on treatment with AmB which has been minimized by the development of lipid formulations of AmB. The formulation targets intramacrophageal organisms, and increases the bioavailability of drugs and reduces their toxicity [9]. Due to their minimal toxicity they can be administered in higher doses over a shorter duration and now it is possible to deliver the total curative dose by single or two infusions. In spite of its safety administration, the uses are beyond the reach to most of the patients in the endemic areas of VL due to affordability [10].

## *Nanoconcept in VL Treatment*

The macrophage phagocytic system (MPS) preferentially engulfs the nanoparticles which is home for the leishmanial parasites ensuring targeted drug delivery [11]. Uniformly and significantly reduced size of drug (1–2 μm to 10–20 nm) increases the contact surface improving drug efficacy [12] Further, nanosizing increases the adhering properties to tissues [13] and has high weight per volume which is an ideal situation for the sustained release as a depot [14]. Such properties minimize the side effects to other systems of the host.

Nanoparticles of AmB, a different presentation of Amphotericin is obtained by "nanonization" [15] which is recognized as foreign bodies and phagocytozed by the MPS [9] allowing targeted drug delivery. We had developed nanoparticles of AmB deoxycholate, the most efficient antileishmanial drug, using high pressure milling homogenization and checked its antileishmanial activities. The in vivo and in vitro experiments have shown promising clearance of promastigotes as well as amastigotes [16–18].

In this study we aimed to evaluate this formulation in the treatment of VL ex vivo by raising infection of J774A cell line by amastigote stage isolated from infected spleen of hamster and raising infection to peritoneal macrophages of hamster by promastigote to compare the responses of natural and clone cell responses to the parasites. Besides, nano-AmB has been traced in vivo to study drug localization using Transmission Electron Microscopy.

# Materials and Methods

## *Animal Model*

Syrian golden hamsters (*Mesocricetus auratus*) (Fig. 1) were selected as animal model to raise infection, which mimic several aspects of human VL [19]. The animal model as reservoir fulfilled the requirement of natural amastigote stage parasites to raise infection ex vivo. Furthermore, the hosts were also used to extract peritoneal macrophage to perform ex vivo experiment and tracking the nanoformulation of drug up to the targeted organ, spleen.

**Fig. 1** Syrian golden hamsters (*Mesocricetus auratus*) selected as animal model for raising infection by promastigotes, extraction of peritoneal macrophages, and tracking the nano-amphotericin drug 7 days after treatment

## *Culture of Leishmania Donovani Parasites and J774A Cell Line*

Splenic aspirate from VL patient was aseptically transferred to NNN tube in endemic district (Muzaffarpur, Bihar, India) for primary culture of *L donovani.* The well-grown parasites in biphasic medium were passaged to monophasic complete RPMI–1640 medium (HyClone) supplemented with 0.2 % $Na_2HCO_3$, 2.05 mM L-Glutamine, 12 mM HEPES buffer (HiMedia, India), 15 % (v/v) heat inactivated FBS (Gibco, Germany), and 50 mg/L Gentamycin at 26 °C. The WHO reference strain of *L. donovani* (MHOM/IN/80/Dd8) conserved in hamsters was used for collection of direct amastigotes to raise infection to cell line. J774A macrophage cell line (obtained from NCCS, Pune) and peritoneal macrophage cells were cultured in the cRPMI–1640 media as above to be used as cellular host for the ex vivo intracellular test of antileishmanial activity against amastigotes and promastigotes, respectively.

## *Isolation of Metacyclic Stage and Raising Ex Vivo Infection*

Isolation of metacyclic stage parasites were performed as described earlier [20] with minor modification. Briefly, 6-day-old parasite culture was centrifuged and washed 2X with 0.02 M PBS (pH 7.2) and resuspended in complete RPMI. Five percent FBS was added to the parasite solution and kept for 30 min. at 37 °C for opsonization in order to isolate the metacyclic stage of parasites. Incubation was followed by 2X washing with PBS before use in ex vivo tests.

## *Conversion of Amphotericin B into Nanoparticles (Nano-Amphotericin B) and Characterization*

The amphotericin B deoxycholate used to treat VL patients was milled using a locally fabricated mechanical miller as described by Manadhar et al. [18]. Briefly, milling was done under argon atmosphere employing 150 atm pressure at a maximum speed of 500 rpm. Three steel balls were used (two steel balls of 12.5 mm diameter and one of 5 mm diameter). The AmB was milled for 30 min followed by sonication for 10 min under benzene.

The milled powder was characterized by scanning (Philips: XL 20) and transmission electron microscope (SEM, Philips E M CM-12, and Tecnai $20G^2$). On SEM, the milled particles were monitored on the stub (SEM specimen holder) with the uniform silver coated film. The specimen preparation for the transmission electron microscopic (TEM) exploration was carried out in a different way. In this case, the sonicated powder was sprayed on distilled water surface in a beaker. These AmB particles, which floated on the water surface were expected to be of thin version. These were mounted on transparent 120 kV electron microscopic copper grids of 300 mesh sizes and were coated with formvar supported grid. Solution of the formvar (0.2 g/100 ml of chloroform) was spread on water and a film was formed on the water surface. Copper specimen grid was put on the formvar film surface. The grid was taken out in such a way that formal film piece covered the grid surface. The AmB particles which floated on the water surface were lifted on the formvar-coated grid. Since Amphotericin is low strength atomic bonded solid, low dose mode of TEM operation was selected to avoid electron beam-induced dissociation.

## *Reconstitution of Drug*

As per the instruction mentioned by the manufacturer, AmB was reconstituted in sterile-chilled triple-distilled water to obtain stock solution of 5 mg/ml. Five percent dextrose was added and stored at 4 °C for maximum a week [21]. Similarly, nano-amphotericin drug was also reconstituted at the same concentration and stored. The drugs were further diluted with RPMI according to the end need of concentration for ex vivo tests in cold chain.

## *Isolation of Amastigote Stage of L. donovani from Infected Spleen of Hamster*

The hamsters infected with *L. donovani* (MHOM/IN/80/Dd8) were biopsized (infected before 30 days) and spleen was taken out aseptically [22]. The infected spleen was homogenized with cRPMI in cold chain and centrifuged at 900 rpm at 4 °C for 5 min. The supernatant was again spun at 2,500 rpm in a similar condition. Then, the supernatant was discarded and the pellet was resuspended in media. The bright shining amastigotes were counted and resuspended so that each 100 μL contained $2.5 \times 10^5$ parasites.

## *Ex Vivo Assay on Intra-J774A Macrophage Amastigotes After Raising Infection Directly by L. donovani Amastigote Isolated from Infected Spleen of Hamster*

J774-A macrophage cell line was used for ex vivo intracellular drug assay. The macrophages were resuspended at $2.5 \times 10^5$ cells per ml in serum-free RPMI–1640. Two hundred microlitre cell-suspensions per well were plated on eight chamber Lab-Tek tissue culture slides (NUNC, USA) and were allowed to adhere for 2 h in $CO_2$ incubator with supply of 5 % $CO_2$ at 37 °C. The wells were washed 2X again with serum-free media and the adherent macrophages were infected with amastigotes parasite suspension of *Leishmania donovani* maintaining leishmania-macrophage ratio of 10:1 in 200 μl final solution of complete media. The chamber slide was incubated for 2 h in the same condition. Free amastigotes were washed out with serum-free medium and the cultures were incubated in the complete medium supplemented with 15 % (v/v) HIFBS (Gibco, Germany) and 50 mg/L gentamycin overnight. After incubation, different concentrations of drugs AmB (0.02–0.00025 μg/ml) and nano-AmB (0.005–0.0006 μg/ml) in duplicate were dispensed except the two control wells (without drug). The slides were then incubated for 72 h followed by methanol fixing for 1 min and stained with Giemsa (Qualigens, Mumbai). The slides were observed for amastigote counting.

## *Ex Vivo Assay on Intraperitoneal Macrophage Amastigotes After Raising Infection by Metacyclic Stage of L. donovani Promastigotes*

Peritoneal lavage from healthy anesthetized hamsters was collected on the next day after peritoneal injection of 2 % starch in aseptic condition. The lavage was centrifuged at 1500 rpm at 10 °C for 15 min followed by 2X washing in similar condition. The macrophages were plated as mentioned above for in vitro assay. The wells with adherent macrophages were infected with parasites of metacyclic stage and left for overnight for infection. The infection procedure was followed by dispensing test drugs nano-AmB (0.005–0.0006 μg/ml) and reference drugs AmB (0.02–0.00025 μg/ml), Miltefosin (0.625–5 μg/ml), and Sodium Stibogluconate (12.5–150 μg/ml) in duplicate except the two control wells. The slides were then incubated for 72 h followed by slide preparation.

## *Calculation of Percentage of Inhibition and $ED_{50}$*

The Giemsa-stained slides of infected macrophage were observed in light microscope (Zeiss, AXIO, Imager A1) and counted the number of amastigotes and infected macrophages per 100 cells. The mean value of count was taken for each control and test wells. Percentage inhibition was calculated by using the formula mentioned below:

$$PI = \frac{PC - PT}{PC} \times 100$$

PI—Percentage of Inhibition, PC—Number of amastigotes per 100 cells in control well, PT—Number of amastigotes per 100 cells in drug-treated well.

## *Nano-AmB Drug Tracking to the Targeted Organ (Spleen)*

An infected hamster (35–45 g) was treated on day-30 post-inoculation (p.i.). Hamsters were anesthetized and operated according to Committee for the Purpose of Control and Supervision of Experiments on Animal (CPCSEA), Delhi, India by injecting thiopentone (Thiosole sodium; Neon Laboratories Limited, India) at 80 mg/kg intraperitoneally. Spleen was removed and minced. TEM was done as mentioned above to trace the drug in the targeted organ.

## *Statistical Analysis*

An unpaired *t* test (two-tailed) was used to determine statistical significance of $ED_{50}$ while comparing the test drug with reference drugs in ex vivo inhibition analysis.

# Results

## *Structure of Nano-Amphotericin B Deoxycholate as Seen by TEM and SEM*

The representative SEM micrographs of the original amphotericin particles were polygonal crystallite with 1–2 μm (Fig. 2a, b).

The TEM micrograph of thin section showed crystalline grains 1 and 2 separated by a grain boundary marked by arrows (Fig. 3a) of original amphotericin particles having size of ~1 μm. The selected area electron diffraction (SAD) pattern brought out the crystalline character of the AmB particles (Fig. 3b). The

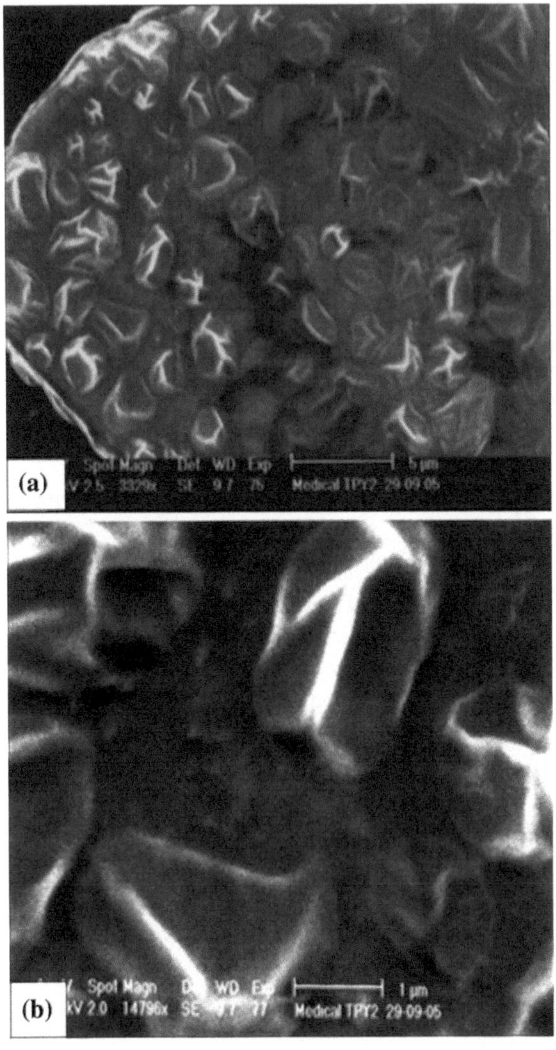


**Fig. 2** Scanning electron micrograph of commercially available Amphotericin B

important relevant result from the SEM and TEM studies illustrated that the AmB particles were quite large (≥1 μm). Therefore, these are massive particles which may produce toxicity.

On high pressure milling the AmB polygonal particles got divided into smaller particles of average particle sizes 10–20 nm (Fig. 4) ranged in nanoregion.

TEM explorations of the milled particles pasteurized the representative results of nanosize formation of AmB (Fig. 5a, b, and c). The bright (Fig. 5a) and dark field version (Fig. 5b) represented TEM picture of the milled nano-amphotericin B. These results confirmed the nano-nature of the milled AmB. The SAD diffraction pattern of the particles of Fig. 5a, b was pictured as in Fig. 5c. This

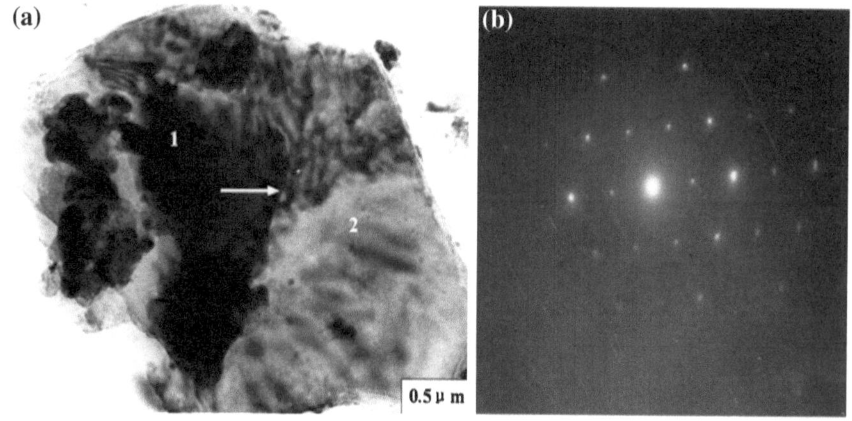


**Fig. 3** **a** Transmission electron micrograph of commercially available original amphotericin B, **b** The selected area electron diffraction pattern bring out the crystalline character of the commercial amphotericin B

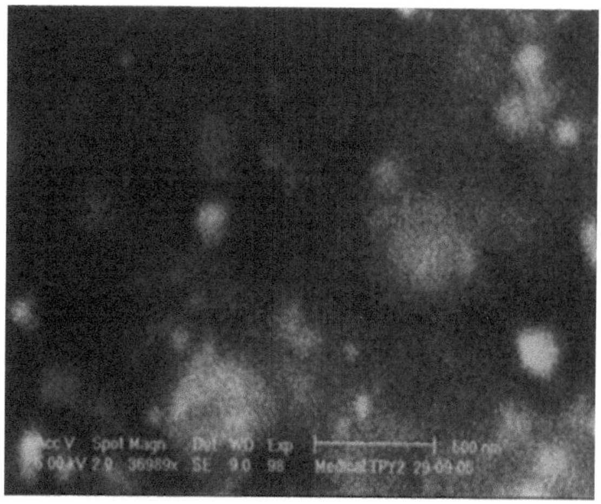

**Fig. 4** Scanning electro micrograph of high pressure milled AmB. The micrograph showed the formation of nano-AmB

diffraction pattern showed nanoparticle characteristics intact even after milling. Hence, the amphotericin B nanoparticles retained its crystalline nature.

Ex vivo **Nano-AmB drug efficacy test on intracellular amastigotes**. The nano-amphotericin drug particles were traced in vivo after 7 days of treatment in homogenized spleen and were found accumulated as a depot (Fig. 6). In ex vivo experiments, the number of amastigotes was counted before treatment and was very high in comparison to the count after treatment for 72 h (Fig. 7).

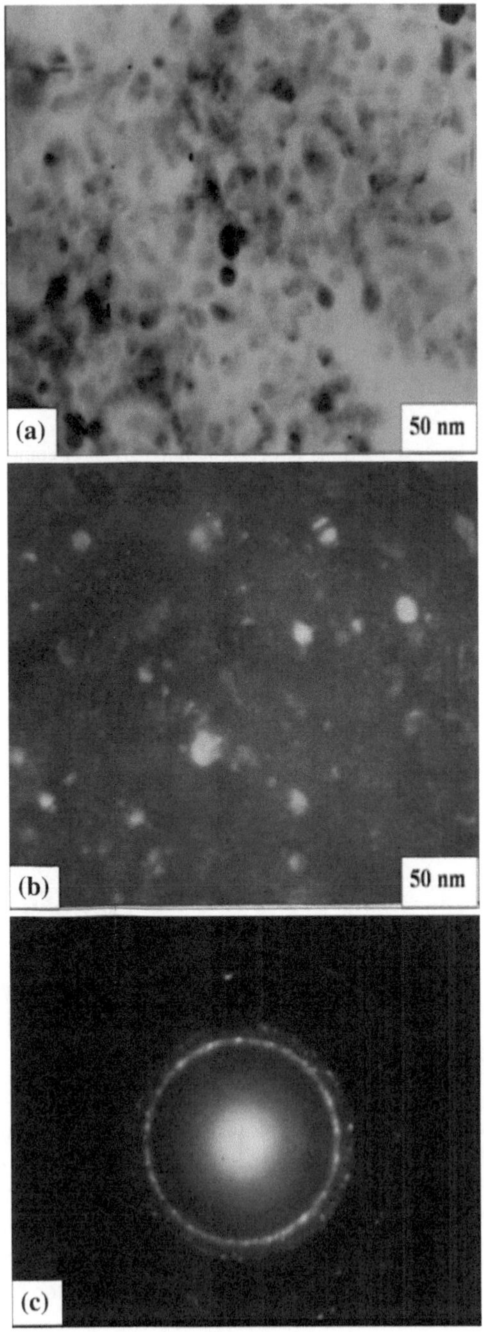


**Fig. 5** **a** Bright field and **b** dark field TEM images of nano-AmB, showing the formation of nanosize of the order of 10–20 nm, **c** the corresponding diffraction ring pattern shows the nanocryatalline phase

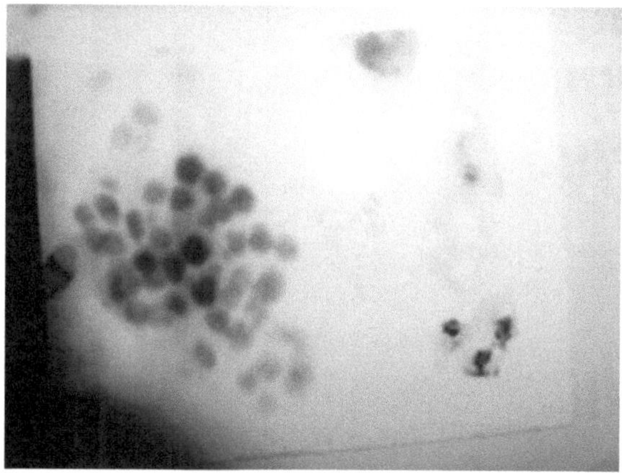

**Fig. 6** Transmission electron microscopy (TEM) of nanoparticles (*dark areas*) accumulated 7 days after treatment with nano-amphotericin B in the homogenized spleen of infected hamster

## *Drug Activity Against Infected Peritoneal Macrophages (Infection Raised by Promastigotes)*

The comparative antileishmanial activities of test drug nano-amphotericin, and reference drugs AmB, miltefosine, and SAG showed that test drug has very high efficacy. The efficacy against the *L donovani* amastigotes in peritoneal macrophage cell line (infection raised by promastigotes) demonstrated significant inhibition by nanoparticles of AmB ($ED_{50}$, 0.0027 $\pm$ 0.001 μg/mL) as compared to conventional AmB ($ED_{50}$, 0.0426 $\pm$ 0.003 μg/mL). The other reference drugs Miltefosin and sodium antimony stibogluconate had ED50 of 1.388 $\pm$ 0.201 and 71.5239 $\pm$ 3.812 (Fig. 8).

## *Drug Activity Against Infected J774A Cell Line (Infection Raised by Amastigotes)*

The host–parasite interaction of virulent amastigote with J774A cell line at its infection phase also demonstrated the similar result of high efficacy of nanoformulation. Fifty percent of intracellular amastigotes were cleared by 0.0038 $\pm$ 0.001 μg/mL while the dose was slightly more for AmB (0.0196 $\pm$ 0.001 μg/mL) which showed the significant difference between nano-AmB and AmB (*p* value, 0.0122) (Table 1).

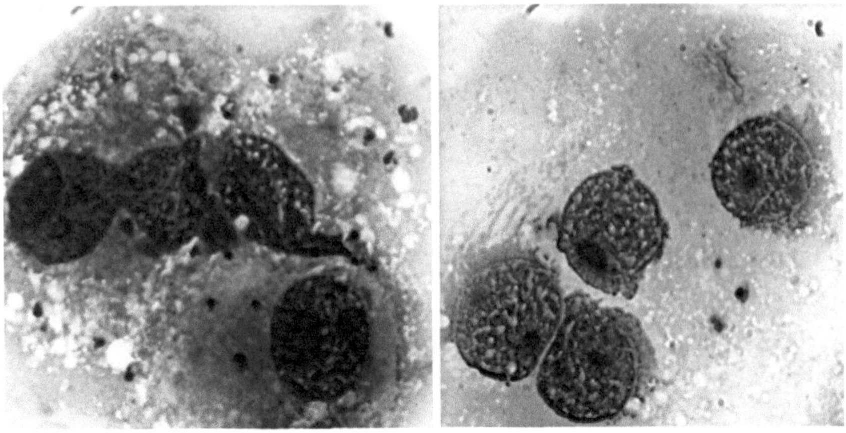

**Fig. 7** Giemsa-stained J774-A macrophage cell line infected by *L donovani*. **a** Before treatment with nano-AmB. **b** After treatment with nano-AmB

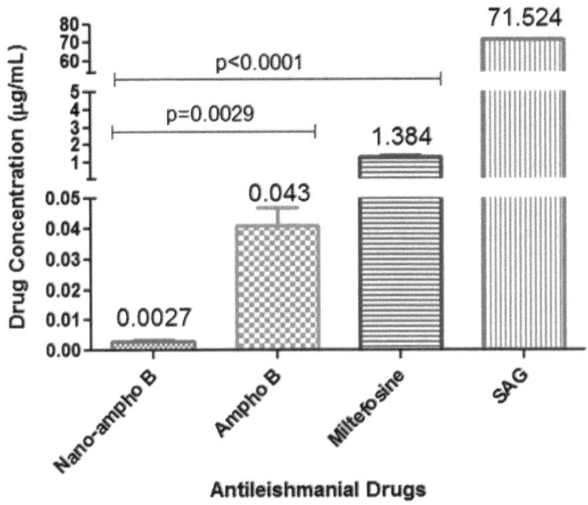


**Fig. 8** Comparative ED50 doses of test drug nano-amphotericin B and reference drugs Amphotericin B deoxycholate, Miltefosin, and Sodium antimony gluconate (SAG) to inhibit intracellular amastigotes

## *Comparative Inhibition of Infected Macrophages by Nano-AmB and AmB*

The inhibited number of infected macrophages after the attempt of treatment also expressed the similar results as of amastigote inhibition. The nanoformulation has shown high efficacy with $ED_{50}$ concentration of 0.0034 ± 0.004 µg/mL for infected peritoneal macrophages infected by promastigote against 0.0369 ± 0.001 µg/mL of AmB. The nano-AmB dose was slightly less to 0.0028 ± 0.001 µg/mL for infected J774 macrophage cell line and AmB was

**Table 1** Drug concentration (μg/mL) for evaluation of efficacy in ex vivo test

| Treatment groups | $ED_{50}$ (mg/L) | | | | |
|---|---|---|---|---|---|
| | Mil | SAG | Nano-AmB | AmB | P-value[a] |
| Amastigotes in peritoneal macrophages (infected by promastigotes)[b] | 1.3844 ± 0.201 | 71.5239 ± 3.812 | 0.0027 ± 0.001 | 0.0426 ± 0.003 | 0.0029 |
| Amastigotes in J774 A cell line (infected by amastigotes of infected spleen of hamster)[b] | ND | ND | 0.0038 ± 0.001 | 0.0197 ± 0.003 | 0.0122 |
| Infected peritoneal macrophages | 3.8125 ± 0.15 | 102.8133 ± 6.01 | 0.0034 ± 0.004 | 0.0369 ± 0.007 | 0.0001 |
| Infected J774 macrophages | ND | ND | 0.0028 ± 0.001 | 0.0181 ± 0.006 | 0.0001 |

[a] P value between nano-AmB and AmB
[b] Number of amastigotes per 100 nuclei, *ND* = not done

slight higher dose 0.0181 ± 0.006 μg/mL for infected J774 cell line infected by virulent amastigotes. The other reference drugs showed ED50 of 3.812 ± 0.15 and 102.813 ± 6.01 by Miltefosin and SAG (Table 1).

## Discussion

Leishmaniasis has been noted as the second widely spreading life-threatening disease after malaria [23]. Many efforts have been implemented for appropriate drug against this disease but still none other than AmB could have the choice of drug in the current situation though it has side effect of toxicity. In light of this, we have given a new face to this drug by converting it into nanoparticles which did not lose their original character. Nevertheless, this 100 % effective drug is now equipped with high weight per volume, which is an ideal situation for the sustained release as a depot [14]. Furthermore, these particles are preferentially engulfed by macrophage cells of MPS which are the targeted cells and localized in spleen, liver, bonemarrow, and lymphnodes. We achieved the basic targeted delivery of the drug by this new formulation as traced in spleen tissues by TEM.

Nano-amphotericin had good efficacy over the AmB in in vitro study on promastigotes [18]. Here, we further evaluated the drug efficacy of the nanoform in comparison to conventional AmB providing different conditions of host–parasite interaction. It is inferred that, nanonization has increased the efficacy of AmB. Nano-AmB is 15.8 times more efficient in reducing the amastigotes' burden in peritoneal macrophages than by AmB in the case peritoneal macrophage infected by promastigote and 5.2 times more efficient in J774 cell line infected by splenic amastigotes (p-value 0.0029 and 0.0122). The differences in $ED_{50}$ in the two host/parasite system might be due to associated host factors from different cell lineages along with the drug kinetics. The comparison in two different host–parasite conditions showed that peritoneal macrophage required 1.4 times low concentration of nano-AmB to clear 50 % amastigotes than AmB. In contrast, $ED_{50}$ in J774A cell line required 2.16 times low concentration of AmB than nano-AmB. The reason behind could be the active ex vivo response of peritoneal macrophage toward nanoparticles. The importance of selecting appropriate macrophage populations when studying pathogen interactions with functionally diverse lineages of cells has already been highlighted by Phillips et al. [24]. However, to discuss the exact role behind the difference in efficacy of the drug to peritoneal and J774A macrophages would be a scope for further study. $ED_{50}$ in inhibition of infected peritoneal/J774 macrophages were also better achieved by nano-AmB than by AmB similar to the trend in growth inhibition of amastigotes supporting the better MPS engulfing nanoparticles (Table 1).

Hence, the newly formulated nano-AmB was more effective to the intramacrophage amastigotes than conventional AmB. Nontoxicity of this formulation has been discussed earlier [18] with CC50 values well beyond the $ED_{50}$ doses (12.67 μg/mL vs. 0.004 μg/mL). The result in the present study suggests that nanoparticles of AmB is superior to commercially available AmB. Thus, smaller doses of nano-AmB may be required to achieve the high cure as observed currently. The wide benefits of nanonization have given the drug with various nanoformulations recently [16–18], but only after their safety and efficacy are closely observed should these be envisioned in any clinical studies.

**Acknowledgments** The authors are thankful to the Nano Science and Technology Initiative, Department of Science and Technology (DST), India for financial support in the nanonization part. We gratefully acknowledge Kala-azar Medical Research Centre, Muzaffarpur, India for providing infected aspirates to culture the parasites. Prof. Manandhar is thankful to UGC, Nepal for providing fellowship to him during the study.

## References

1. Ashford RW, Desjeux P, Deraadt P (1992) Estimation of population at risk of infection and number of cases of leishmaniasis. Parasitol Today 8:104–105
2. Desjeux P (2004) Leishmaniasis: current situation and new perspectives. Comp Immunol Microbiol Infect Dis 27:305–318
3. Bora D (1999) Epidemiology of visceral leishmaniasis in India. Natl Med J India 12:62–68
4. Lal S, Saxena NBL, Dhillon GPS (1996) Kala-azar cases and deaths. In: manual on visceral leishmaniasis (kala-azar) in India. National Malarial Eradication Programme, Annexure VII, New Delhi, pp 167-77
5. Berman JD (1997) Human leishmaniasis: clinical, diagnostic, and chemotherapeutic developments in the last 10 years. Clin Infect Dis 24:684–703
6. Sundar S, More DK, Singh MK, Singh VP, Sharma S, Makharia A, Kumar PC, Murray HW (2000) Failure of pentavalent antimony in visceral leishmaniasis in India: report from the center of the Indian epidemic. Clin Infect Dis 31:1104–1107

7. Thakur CP, Narayan S, Ranjan A (2004) Epidemiological, clinical and pharmacological study of antimony-resistant visceral leishmaniasis in Bihar, India. Indian J Med Res 120:166–172
8. Sundar S, Gupta LB, Rastogi V, Agrawal G, Murray HW (2000) Short-course, cost-effective treatment with amphotericin B-fat emulsion cures visceral leishmaniasis. Trans R Soc Trop Med Hyg 94:200–204
9. Muller RH, Jacobs C, Kayser O (2001) Nanoparticles as particulate drug formulations in therapy rationale for development and what we can expect for future. Adv Drug Deliv Rev 47:3–19
10. Murray HW (2001) Clinical and experimental advances in treatment of visceral leishmaniasis. Antimicrob Agents Chemother 45:2185–2197
11. Muller K, van Zandbergen G, Hansen B, Laufs H, Jahnke N, Solbach W, Laskay T (2001) Chemokines, natural killer cells and granulocytes in the early course of Leishmania major infection in mice. Med Microbiol Immunol 190:73–76
12. Calka A, Wexler D (2001) Mechanical milling assited by electrical discharge. Nature 419:147–151
13. Yoshiaki K (2001) Nanoparticulate systems for improved drug delivery. Adv Drug Deliv Rev 47:1–2
14. Kayser O, Lemke A, Hernandez-Trejo N (2005) The impact of nanobiotechnology on the development of new drug delivery systems. Curr Pharm Biotechnol 6:3–5
15. Rabinow BE (2004) Nanosuspensions in drug delivery. Nat Rev Drug Discovery 3:785–796
16. Prajapati VK, Awasthi K, Yadav TP, Rai M, Srivastava ON, Sundar S (2011) An oral formulation of amphotericin B attached to functionalized carbon nanotubes is an effective treatment for experimental visceral leishmaniasis. J Infect Dis 205:333–336
17. Prajapati VK, Awasthi K, Gautam S, Yadav TP, Rai M, Srivastava ON, Sundar S (2011) Targeted killing of Leishmania donovani in vivo and in vitro with amphotericin B attached to functionalized carbon nanotubes. J Antimicrob Chemother 66:874–879
18. Manandhar KD, Yadav TP, Prajapati VK, Kumar S, Rai M, Dube A, Srivastava ON, Sundar S (2008) Antileishmanial activity of nano-amphotericin B deoxycholate. J Antimicrob Chemother 62:376–380
19. Bunn-Moreno MM, Madeira ED, Miller K, Menezes JA, Campos-Neto A (1985) Hypergammaglobulinaemia in Leishmania donovani infected hamsters: possible association with a polyclonal activator of B cells and with suppression of T cell function. Clin Exp Immunol 59:427–434
20. Franke ED, McGreevy PB, Katz SP, Sacks DL (1985) Growth cycle-dependent generation of complement-resistant Leishmania promastigotes. J Immunol 134:2713–2718
21. Yardley V, Croft SL (2000) A comparison of the activities of three amphotericin B lipid formulations against experimental visceral and cutaneous leishmaniasis. Int J Antimicrob Agents 13:243–248
22. Bhatnagar S, Guru PY, Katiyar JC, Srivastava R, Mukherjee A, Akhtar MS, Seth M, Bhaduri AP (1989) Exploration of antileishmanial activity in heterocycles; results of their in vivo and in vitro bioevaluations. Indian J Med Res 89:439–444
23. ScienceDaily (2011) New treatment for Kala Azar, the most deadly parasitic disease after malaria. http://www.sciencedaily.com/releases/2011/09/110923102525.htm. Accessed 23 Mar 2012
24. Phillips R, Svensson M, Aziz N, Maroof A, Brown N, Beattie L, Signoret N, Kaye PM (2010) Innate killing of Leishmania donovani by macrophages of the splenic marginal zone requires IRF-7. PLoS Pathog 6:e1000813

# Editors Profile

**Dr. Rameshwar Adhikari** is Associate Professor at Central Department of Chemistry, Tribhuvan University, Kathmandu. He has been Invited Professor in Rouen University (France), Mahatma Gandhi University (India), and Tokyo Institute of Technology (Japan). He is member of Nepal Chemical Society and German Engineers Society. He has been bestowed several awards including Ambassador Scientist of Alexander von Humboldt Foundation, Fellowship of International Union of Pure, and Applied Chemistry and POLYCHAR World Forum International Materials Science Prize. He is President of Nepal Polymer Institute and has convened several International Congresses, Symposia, and Workshops. His research areas include nanomaterials, biopolymers, and biomedical materials. He has authored over 100 research papers in peer-reviewed journals, co-authored four textbooks, and edited two books.

**Mr. Santosh Thapa** completed his M.Sc. (Medical Microbiology) from Tribhuvan University, Kathmandu and taught related courses to undergraduate students at departments of microbiology of Tribhuvan University and affiliated colleges. He is the founder General Secretary of Nepalese Association of Medical Microbiology, members of American Society for Microbiology and Society for Applied Microbiology, and associated with International Union of Microbiological Societies. He has worked toward organizing several international conferences and workshops and has authored five research papers in peer-reviewed journals. Currently, he is at Graduate School of Biomedical Sciences, University of North Texas Health Science Center, Fort Worth, Texas, USA for his Ph.D. study.

R. Adhikari and S. Thapa (eds.), *Infectious Diseases and Nanomedicine II*,
Advances in Experimental Medicine and Biology 808,
DOI: 10.1007/978-81-322-1774-9, © Springer India 2014

# Author Index




R. Adhikari and S. Thapa (eds.), *Infectious Diseases and Nanomedicine II*,
Advances in Experimental Medicine and Biology 808,
DOI: 10.1007/978-81-322-1774-9, 

# Subject Index



R. Adhikari and S. Thapa (eds.), *Infectious Diseases and Nanomedicine II*, Advances in Experimental Medicine and Biology 808, DOI: 10.1007/978-81-322-1774-9, © Springer India 2014